HYDROLOGICAL DROUGHT FORECASTING IN AFRICA AT DIFFERENT SPATIAL AND TEMPORAL SCALES

Cover page: Images of different drought events.

HYDROLOGICAL DROUGHT FORECASTING IN AFRICA AT DIFFERENT SPATIAL AND TEMPORAL SCALES

DISSERTATION

Submitted in fulfillment of the requirements of
the Board for Doctorates of Delft University of Technology
and
of the Academic Board of the UNESCO-IHE
Institute for Water Education
for
the Degree of DOCTOR
to be defended in public on
Tuesday, 23 June 2015, 12:30 PM
In Delft, the Netherlands

by

Patricia M. TRAMBAUER ARECHAVALETA

Master of Science in Water Science and Engineering
UNESCO-IHE Institute for Water Education

born in Montevideo, Uruguay

CRC Press
Taylor & Francis Group
Boca Raton London New York

CRC Press is an imprint of the
Taylor & Francis Group, an informa business
A BALKEMA BOOK

This dissertation has been approved by the
promotor: Prof. dr. S. Uhlenbrook and
copromotor: Dr. S. Maskey

Composition of the Doctoral Committee:

Chairman	Rector Magnificus, Delft University of Technology
Vice-Chairman	Rector UNESCO-IHE
Prof. dr. S. Uhlenbrook	UNESCO-IHE / Delft University of Technology, promotor
Dr. S. Maskey	UNESCO-IHE, copromotor

Independent members:

Prof. dr. M. Bierkens	Utrecht University
Prof. dr. W. Bastiaanssen	Delft University of Technology / UNESCO-IHE
Prof. dr. M. McClain	UNESCO-IHE / Delft University of Technology
Dr. R. Teuling	Wageningen University
Prof. dr. C. Zevenbergen	Delft University of Technology / UNESCO-IHE, reserve member

This research was conducted under the auspices of the Graduate School for Socio-Economic and Natural Sciences of the Environment (SENSE)

First issued in hardback 2018

CRC Press/Balkema is an imprint of the Taylor & Francis Group, an informa business

Published by:
CRC Press/Balkema
PO Box 11320, 2301 EH Leiden, The Netherlands
e-mail: Pub.NL@taylorandfrancis.com
www.crcpress.com – www.taylorandfrancis.com

ISBN 13: 978-1-138-37336-5 (hbk)
ISBN 13: 978-1-138-02865-4 (pbk)

Acknowledgments

First and foremost, I would like to express my deepest gratitude to my supervisory team: Prof. Stefan Uhlenbrook, Dr. Shreedhar Maskey and Dr. Micha Werner. Their guidance and scientific advice was invaluable and enabled me to develop the subject and to grow as a researcher. Stefan, thanks for the insightful discussions and accurate suggestions. Shreedhar, I am heartily thankful for your admirable guidance, trust in my skills, patience, constant support and encouragement throughout the whole research period. Micha, you are one of the most enthusiastic and energetic persons that I know and I admire the passion that you have for your work. I am forever grateful for transmitting that motivation to me, for your exhaustive guidance and assistance, for our extensive talks and discussions, and for sharing fascinating and memorable trips with me along the way.

I gratefully acknowledge the funding source of my PhD, the DEWFORA (Improved Drought Early Warning and FORecasting to strengthen preparedness and adaptation to droughts in Africa) project which was funded by the Seventh Framework Programme for Research and Technological Development (FP7) of the European Union (Grant agreement no: 265454). This thesis was part of the DEWFORA project in collaboration with 18 partners, 8 from Europe and 10 from Africa. Being part of this project was an amazing experience that gave me the opportunity to work with a great team of experts and amazing people. The contribution of a number of colleagues involved in the project was crucial for the completion of this work. Special thanks to Dr. Florian Pappenberger, Dr. Emanuel Dutra, Dr. Hessel Winsemius, Mathias Seibert and Dr. Ilyas Masih for their invaluable support and fruitful discussions during this period.

My sincere gratitude goes to Dr. Rens van Beek for sharing the PCR-GLOBWB hydrological model with me and for his assistance and helpful discussions. My thanks go to Dr. Francois Engelbrecht and Dr. Willem Landman from CSIR who hosted me for a couple of months in Pretoria, South Africa.

My time in The Netherlands was greatly enjoyable mostly due to the many friends that became a part of my life and grew into my family abroad. Vero, thanks for being such an amazing friend, for your sincere kindness and beautiful humour. Saul, my adopted brother, your friendship is precious. Ana, you enriched my first years away from home and your friendship remains intact despite the distance. My dearest friends Assyieh, Ruski, Ina, Andreas, Maria, Mulo, Davide, Guy, Maribel, Angy, Neiler, Juliette, Yared, Gonza, Juanca, Joana, Fer, Jessi, Mauri, Erika, Aline, Jelstje, Sol, thank you for being there for all these years. And to all my friends in Uruguay, thanks for always remaining by my side and making me feel very near in spite of the distance. Your friendship is priceless.

Last but not least, I would like to thank my family for all their love, encouragement and belief in me. Words cannot express how grateful I am to my parents, for blindly supporting me in all my pursuits. To my brothers, for their wise advises, always taking care of me and giving me the opportunity to finally have beautiful sisters, which I could not have chosen better. Special thanks to my grandfather, for his interest in my work that led to motivating talks and discussions. And to Aki, for being my source of inspiration and showing me that anything is possible. Thanks for your love, patience and faithful support. It would have been impossible to go through all these years away from home without you. Thank you!

"Be the change that you wish to see in the world."— *Mahatma Gandhi*

Summary

Africa has been severely affected by droughts in the past contributing to food insecure conditions in several African countries. Recent studies show that the frequency and severity of droughts seems to be increasing in some areas as a result of climate variability and climate change. Moreover, the rapid increase of population will certainly aggravate water shortage at local and regional scales. In view of the severe drought conditions and water shortage that may be expected in the coming years in sub-Saharan Africa, efforts should focus on improving drought management by ameliorating resilience and preparedness to drought.

The main objective of this research is to contribute to the development of a modelling framework for hydrological drought forecasting in sub-Saharan Africa as a step towards an effective early warning system. With this aim, a number of specific milestones were defined. These include (i) selecting a proper hydrological model, (ii) testing the model on a continental scale, (iii) assessing the performance of the model in characterising past droughts on a regional scale, (iv) examining the potential of downscaling the low resolution hydrological model results to the high resolution grid, and (v) assessing the skill of a hydrological forecasting system.

The modelling framework developed followed a sequence of defined steps. A set of criteria were defined to assess the suitability of hydrological models for drought forecasting. A number of models were then assessed, and from the subset of the suitable models, a model was selected and applied for the entire African continent. The resulting actual evaporation from the model was analysed and compared with other independently computed evaporation products for different geographical and climatic regions of Africa. A higher resolution version of the same hydrological model was then applied to the water-stressed Limpopo River basin in southern Africa, and the performance of the model in simulating space-time variability of past reported droughts in the basin was analysed. For this analysis, different meteorological, agricultural and hydrological drought indicators were computed and used to identify and characterise past droughts and their severity. The effect of spatial resolutions on three distributed fluxes or storages, i.e. actual evaporation, soil moisture, and total runoff, was computed by using the results of both the high ($0.05° \times 0.05°$) and low ($0.5° \times 0.5°$) resolution models. The coefficient of variation was used to assess the variability of the high resolution variables within each low resolution pixel. Two different techniques were then applied to downscale hydrological variables; i.e. bias correction statistical downscaling, and downscaling by using topographic and soil attributes. Lastly, the high resolution hydrological model was used to set up and test three probabilistic seasonal forecasting systems for the Limpopo Basin, which were forced with different meteorological ensemble forecasts. The skill of the forecasting systems in predicting streamflow and other useful drought indicators was assessed with standard verification skill scores.

Among the 16 well-known hydrological and land surface models selected for the review, five showed good potential for hydrological drought forecasting in Africa. From this subset of models, the PCR-GLOBWB hydrological model was selected and applied for the entire African continent. From the evaporation analysis, results provided a range in actual evaporation that can be expected in a given region in Africa. Moreover, an Actual Evaporation Multiproduct at a 0.5° resolution was derived. This evaporation multiproduct integrates satellite based products, evaporation results from land-surface models and from hydrological models forced with different precipitation and potential evaporation data sets, and may serve as a reference data set (benchmark) for Africa. Results from the same hydrological model but applied at a higher resolution over the Limpopo River basin showed that the model is able to represent the most severe droughts in the basin and to identify the spatial variability of past droughts. The analysis revealed that even though meteorological indicators with different aggregation periods serve to characterise different types of droughts reasonably well, there is an added value in computing indicators based on the hydrological model for the identification of droughts and their severity. The analysis of the variability of the fluxes or storages on high resolution grid cells under different land features and soils indicated that there is good potential of downscaling the low resolution hydrological model results to high resolution based only on the terrain and soil characteristics. Finally, the proposed seasonal forecasting system forced with meteorological forecasts from a global atmospheric model (ECMWF seasonal forecast system S4) was found to be skilful in predicting hydrological droughts during the summer rainy season. Results showed that in the Limpopo Basin, the persistence of the initial hydrological conditions contribute to the predictability up to 2 to 4 months, while for longer lead times the predictability of the system is dominated by the meteorological forcing. A simpler forecasting system that is forced with resampled historical data but conditioned on the El Niño Southern Oscillation (ENSO) signal also showed good potential for seasonal hydrological drought forecasting in the Limpopo River basin.

This research contributed to the development of a hydrological drought forecasting system in the African continent at different time horizons and spatial scales. Results of this research contributed to a course on drought forecasting which aimed to transfer the knowledge developed to practitioners and develop capacity in Africa and other regions. An improved forecast capability reinforces an effective early warning system, and an effective drought forecasting and warning system will hopefully contribute to important aspects in the region such as water security, food security, hazard management, and risk reduction.

Samenvatting

Afrika is in het verleden regelmatig zwaar getroffen door droogte, die in veel gevallen hebben bijgedragen aan voedseltekorten in de verschillende Afrikaanse landen. Recente studies tonen aan dat in sommige gebieden de frequentie en de ernst van dergelijke droogte gebeurtenissen lijken toe te nemen als gevolg van klimaatvariatie en klimaatverandering. Bovendien, zal de snelle toename van de bevolking de watertekorten op lokale en regionale schaal verergeren. Gezien de verwachting van het vaker voorkomen van ernstige droogte en watertekorten, zullen we ons sterk moeten richten op de verbetering van droogtebeheer, onder andere door het verbeteren van de weerbaarheid tegen en het voorbereid zijn op droogte.

De belangrijkste doelstelling van dit onderzoek is om bij te dragen aan de ontwikkeling van een framework voor het voorspellen van hydrologische droogte in zuidelijk Afrika, als een stap op weg naar een doeltreffend systeem voor vroegtijdige waarschuwing van droogte. Om dit doel te bereiken zijn een aantal specifieke mijlpalen gedefinieerd. Deze omvatten: (i) het selecteren van een adequaat hydrologisch model, (ii) het testen van dat model op continentale schaal, (iii) het beoordelen van de prestaties van het model in het karakteriseren van historisch opgetreden droogteperioden op regionale schaal, (iv) het onderzoeken van de mogelijkheden om de lage resolutie van de resultaten van het hydrologische model te verfijnen naar een hogere resolutie, en (v) het beoordelen van de nauwkeurigheid van de vervaardigde hydrologische voorspellingen.

De ontwikkeling van deze model-framework volgde een reeks (van te voren) gedefinieerde stappen. Een aantal criteria werden gedefinieerd om de geschiktheid van de hydrologische modellen om droogte te kunnen voorspellen te beoordelen. Een reeks modellen werd vervolgens op deze criteria beoordeeld, en uit de sub-set van geschikte modellen, werd een model gekozen en toegepast op het hele Afrikaanse continent. Een van de belangrijke modelresultaten is de actuele verdamping, die werd geanalyseerd en vergeleken met andere onafhankelijk berekende verdampingsproducten voor verschillende geografische en klimatologische regio's van Afrika. Een hogere resolutie versie van hetzelfde hydrologische model werd vervolgens toegepast op het stroomgebied van de Limpopo Rivier in zuidelijk Afrika. Deze rivier heeft een semi-aride klimaat, en er is vaak sprake van extreme waterschaarste. De prestaties van het model om de variabiliteit van historisch opgetreden droogtes in tijd en ruimte te simuleren werd geanalyseerd. Hiervoor werden verscheidene meteorologische, agrarische- en hydrologische droogte indicatoren berekend en toegepast om de ernst van die eerdere droogtes te karakteriseren. De invloed van de ruimtelijke resolutie op drie gedistribueerde parameters; actuele verdamping, bodemvocht en de afvoer, is berekend met behulp van de resultaten van zowel een model met een hoge (0,05° x 0,05°) en een lage (0,5° x 0,5°) resolutie. Hierbij is de variatiecoëfficiënt gebruikt om de variabiliteit van de hoge resolutie resultaten binnen elke lage resolutie pixel te beoordelen.

Twee technieken werden vervolgens toegepast om de schaal van de hydrologische variabelen te verkleinen: een statistische correctiemethode, en een methode waarbij de schaalverkleining is afgeleid uit hogere resolutie topografische- en bodem eigenschappen. Tot slot is het hoge resolutie hydrologisch model gebruikt om drie probabilistische voorspellingsystemen voor het stroomgebied van de Limpopo op te zetten en te testen voor het maken van voorspellingen op seizoenschaal. De drie systemen maken gebruik van hetzelfde hydrologisch model, maar verschillen van elkaar door de verschillende meteorologische voorspellingen die als randvoorwaarde worden gebruikt. De betrouwbaarheid waarmee deze drie systemen afvoer en andere nuttige droogte indicatoren kan voorspellen is geëvalueerd met een aantal standaard verificatie scores.

Onder de 16 geselecteerde hydrologische en land-oppervlakte modellen die zijn beoordeeld, toonden vijf goede mogelijkheden voor het voorspellen van hydrologische droogte in Afrika. Uit deze deelverzameling van modellen, werd het hydrologische model PCR-GLOBWB geselecteerd en toegepast voor het gehele Afrikaanse continent. Uit de resultaten verkregen van de analyse van de verdamping, werd het bereik van de actuele verdamping afgeleid dat kan worden verwacht in de verschillende regio's in Afrika afgeleid. Op basis hiervan is een verdampingsproduct ontwikkeld, samengesteld uit de actuele verdamping zoals berekend door de verschillende modellen. Dit verdampingsproduct integreert satelliet-gebaseerde producten, resultaten van land-oppervlakte modellen en hydrologische modellen met verschillende neerslag en potentiële verdamping data sets, en kan gebruikt worden als een referentie voor Afrika. Resultaten van het gekozen hydrologische model, maar toegepast op een hogere resolutie over het stroomgebied van de Limpopo toonden aan dat het model goed in staat is de meest ernstige droogte in het stroomgebied aan te tonen, en de ruimtelijke variabiliteit van de historisch opgetreden droogtes juist te identificeren. Uit het onderzoek bleek dat ondanks dat meteorologische droogte-indicatoren met verschillende aggregatie perioden de verschillende typen droogte redelijk goed kunnen karakteriseren, er wel degelijk toegevoegde waarde is in het berekenen van indicatoren op basis van de resultaten van het hydrologische model voor de identificatie de droogte en de ernst daarvan. De analyse van de variabiliteit van de fluxen van de hoge resolutie raster cellen onder verschillende topografische- en bodem eigenschappen gaven aan dat er goede mogelijkheden zijn voor het verkleinen van de schaal van de lage resolutie hydrologische modelresultaten naar hogere resolutie, slechts gebruik makend van die topografische- en bodem eigenschappen. Tot slot bleek het voorgestelde seizoens-voorspellingsysteem, gebruik makend van een globaal atmosferische model-voorspelling (ECMWF S4 seizoenvoorspellingsyteem.) in staat te zijn hydrologische droogte tijdens het regenseizoen in de zomer te voorspellen. Resultaten laten zien dat voor het stroomgebied van de Limpopo de persistentie van de initiële hydrologische condities tot 2 tot 4 maanden doorwerken en dus aanzienlijk bijdragen aan de voorspelbaarheid van hydrologische droogte. Voor voorspellingen met een langere zichttijd wordt de voorspelbaarheid door de ingevoerde meteorologische randvoorwaarden bepaald. Een eenvoudiger voorspellingsysteem, waarbij historische gegevens, geconditioneerd op het ENSO signaal, worden gebruikt als meteorologische randvoorwaarden, bleek ook goede potentie te hebben voor het vervaardigen van hydrologische droogte voorspellingen op seizoenschaal in het stroomgebied van de Limpopo rivier.

Dit onderzoek heeft bijgedragen aan de ontwikkeling van een hydrologisch droogte voorspellingsysteem op verschillende tijdshorizonten en ruimtelijke schalen voor het Afrikaanse continent. Resultaten van dit onderzoek hebben bijgedragen aan de ontwikkeling van een cursus om de vergaarde kennis over te dragen aan vakmensen in Afrika en andere regio's, om zo de capaciteit in deze landen te ontwikkelen. Een verbeterd vermogen tot het maken van hydrologische droogtevoorspellingen zoals in deze studie is aangetoond versterkt een doeltreffend systeem voor vroegtijdige waarschuwing, en een effectief droogtevoorspelling- en waarschuwingssysteem zal hopelijk bijdragen aan belangrijke aspecten in de regio, zoals: water veiligheid, voedselzekerheid, rampenbeheer, en risicovermindering.

Contents

1

GENERAL INTRODUCTION

This chapter is partially based on:

Maskey, S., and Trambauer, P.: Hydrological modeling for drought assessment, in: Hydro-Meteorological Hazards, Risks, and Disasters, 1 ed., edited by: Shroder, J. F., Paron, P., and Di Baldassarre, G., Elsevier, 263-282, 2014, and

Masih, I., Maskey, S., Mussá, F. E. F., and Trambauer, P.: A review of droughts on the African continent: a geospatial and long-term perspective, Hydrol. Earth Syst. Sci., 18, 3635-3649, doi: 10.5194/hess-18-3635-2014, 2014.

1.1 Background

1.1.1 Droughts

Drought is normally defined as a prolonged period of abnormally dry weather condition leading to a severe shortage of water. According to the American Meteorological Society (AMS, 1997), droughts originate from a deficiency of precipitation resulting in water shortage for some activity, and its severity may be aggravated by other meteorological elements. Drought is a normal, recurring feature of climate and it occurs in virtually all climatic regimes. While aridity is a permanent feature of a regional climate, drought is a temporary deviation from a normal condition. Thus, drought should be considered relative to some long-term average condition of balance between precipitation and evaporation in a particular area (AMS, 1997).

Even though drought is a normal feature of climate, it is also a widespread natural hazard with large socio-economic and environmental impacts. Drought develops quite slowly, and its effect may last for several months even after the drought is over. The spatial extent of droughts is normally more widespread than other natural hazards such as floods. Records of the past several decades show that droughts have resulted in a higher number of casualties than all the other hazards combined (EM-DAT, 2014). According to the "Emergency Events Database (EM-DAT, 2014) - the International Disaster Database" of the Centre for Research on the Epidemiology of Disasters, at least 642 drought events were reported across the world from 1900 to 2013. The estimated death toll of these events is about 12 million, while the number of affected people is more than 2 billion, and economic damages were over US $135 billion (Masih et al., 2014). The impacts of drought go beyond the loss of life and property. For example, Alston and Kent (2004) state that intangible damages such as impacts on environment and ecosystems are mostly unaccounted for; repeated droughts can advance an area into desertification (a process of land degradation with severe and irreversible loss of productivity); severe depletion of groundwater tables can have far-reaching consequences (e.g., a long-term reduction in streamflows particularly during dry seasons) and may require many years to decades to replenish. Similarly, other environmental impacts include disappearance of small lakes, wetlands and springs; loss of vegetation; loss of nutrients; soil erosion; increased vulnerability to forest fires; etc. Social impacts can include conflicts and wars, increase of water-borne diseases, and migration or relocation (West, 2014).

Undoubtedly, the impacts of drought are very complex. They cover many sectors of the economy and reach well beyond the area experiencing physical drought. This complexity exists because water is essential to society's ability to produce goods and provide services (Ferrer et al., 2008). However, there are also a number of positive aspects of droughts, and while these are limited they are often overlooked. Drought plays a beneficial role in rejuvenating wetlands. When a wetland dries out and sediments are exposed to the air, the oxygen reinvigorates decomposition, nutrients are released, and dormant plant seeds have a chance to germinate and grow (Brakhage, 2008). In some cases, the production of some nuts and winery grapes has also improved due to drought conditions. Moreover, the interactions across regions can be very intricate. For instance, a big drought in Brazil affecting the coffee production can (positively) affect the economy of other coffee producing countries such as Colombia and Ethiopia. Despite there being some positive effects of droughts, the overall impact in a given area nearly always remains negative.

Recent studies show that the frequency and severity of droughts seems to be increasing in some areas as a result of climate variability and climate change (IPCC, 2007b; Patz et al., 2005; Sheffield and Wood, 2008; Lehner et al., 2006). Africa has been severely affected by droughts in the past contributing to food insecure conditions in several African countries. The United Nations Development Programme (UNDP) estimated that around 220 million people were found to be exposed annually to drought and African states were indicated as having the highest vulnerability to drought. They also state that translation of drought into famine is influenced by armed conflict, internal displacement, HIV/AIDS, poor governance and economic crisis (UNDP, 2004). Moreover, the rapid increase of world population will certainly aggravate water shortage at local and regional scale. The UN (2004) reports that population growth will be much faster in Africa than in other regions of the world, which will add one billion people to the continent, and rise from 13 to 20 percent of world population by 2050.

1.1.2 Droughts in Africa

This section presents a brief review of droughts in Africa, which is largely based on a more extensive review presented in Masih et al. (2014). The frequency, intensity and spatial coverage of droughts have significantly increased across the whole African continent during the second half of the period from 1900-2013. This is supported by studies conducted at continental scale (e.g. Dai, 2011, 2013) as well as by regional and country level studies (Touchan et al., 2011; Elagib and Elhag, 2011; Kasei et al., 2010; Richard et al., 2001). The available data (though limited in temporal coverage) from EM-DAT also supports this observation. This is further illustrated in Figure 1-1, which shows the fraction of the African continent under different drought categories based on an analysis of the Standardized Precipitation Evaporation Index (SPEI) data (http://sac.csic.es/spei/database.html). The widely used non-parametric Spearman Rank test was applied to test the statistical significance of trends in these data, showing a statistically significant increase (at the 99% significance level) in the area under all categories of drought (e.g. moderate, severe and extreme droughts) for the African continent during 1901–2011.

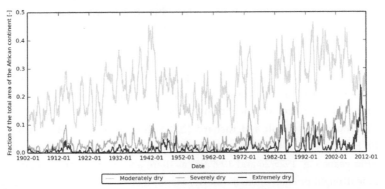

Figure 1-1 Fraction of the African continent under different drought conditions computed from the 12 month SPEI dataset (Source: Global SPEI database available at http://sac.csic.es/spei/database.html, version 2.2 retrieved in May 2014). Note: Moderate dry (-1.5 < SPEI ≤-0.5), Severely dry (-2.0 < SPEI ≤ -1.5), and Extremely dry (SPEI ≤-2.0).

There are few studies available to date that offer possibility of comparing droughts observed during 1900-2013 with those witnessed before the 20th century. Such a comparison is important to show the decadal and century-scale variability in Africa. The work of Touchan et al. (2008, 2011) presents a long term perspective on droughts in Northwest Africa using tree-ring records. They show that the frequency of the occurrence of a one-year duration drought (with a PDSI < -1.05) was 12 to 16 times per century before the 20th century, which increased to 19 during the 20th century. The latter half of the 20th century is seen as the driest period in the last nine centuries. This shift to drier conditions is attributed to anthropogenic climate change.

Several researchers studied historic droughts in Africa based on lake sediment analyses. Shanahan et al. (2009) found that the severe droughts in the Sahel in recent decades are not anomalous compared to the droughts of the past three millennia and indicate that longer and more severe droughts can be expected. In equatorial eastern Africa Verschuren et al. (2000) found the period 1000 to 1270 to be the driest over the last 1100 years and Bessems et al. (2008) noted extreme droughts in the region some 200 years ago. Endfield and Nash (2002) describe the discourse on long-term desertification of the African continent, which emerged during 19th century. Their study shows that major multi-year droughts in southern Africa occurred in 1820-7, 1831-5, 1844-51, 1857-65, 1877-86 and 1894-99. Recent researches suggest that even more severe medium-term (approximately 9 months) droughts can be expected in central and southern Africa (Dai, 2011, 2013). On the contrary, the Sahel region may receive more rainfall, though large uncertainties exist in these findings (Druyan, 2011).

Drought is part of a natural climatic variability, which is quite high at intra-annual, interannual, decadal and century timescales in the African continent. Many studies investigate the natural causes that could be associated with droughts in Africa (Caminade and Terray, 2010; Dai, 2011, 2013; Manatsa et al., 2008; Lebel et al., 2009; Zeng, 2003; Shanahan et al., 2009; Tierney et al., 2013; Vicente-Serrano et al., 2012; Richard et al., 2001; Giannini et al., 2008; Hastenrath et al., 2007; Herweijer and Seager, 2008). Some of these also focus on anthropogenic factors, such as climate change, aerosol emissions, land use practices and resulting land–atmosphere interactions, inducing drought mechanisms (Dai, 2013, 2011; Hwang et al., 2013; Lebel et al., 2009; Zeng, 2003). These studies indicate a number of factors that contribute to drought conditions. However, despite regional differences in the causes of droughts in a specific region, El Niño–Southern Oscillation (ENSO) and sea surface temperatures (SSTs) are regarded as major influencing factors across the continent.

Droughts in southern Africa occur mostly during the warm phase of ENSO. Nicholson and Kim (1997) studied the correlation between precipitation over Africa and ENSO in the Pacific, and found that among the 20 extreme rainfall events they analyzed, 15 events appear to be modulated by ENSO. Additionally, Rouault and Richard (2005) show that 8 out of 12 droughts detected coincide with El Niño years. However, some studies point out that droughts do not always occur during El Niño years (e.g. 1925/26 and 1997/98) as there are many other local and global factors influencing the drought phenomenon. Similarly, Manatsa et al. (2008) suggests that El Niño alone is not a sufficient predictor of droughts in southern Africa. They recommend that the extreme positive sea level pressure (SLP) anomalies at Darwin, Australia, for the averaged March to June period (MAMJ Darwin) are an ideal additional candidate for drought monitoring and forecasting in Zimbabwe and southern Africa.

Contrary to southern Africa, regions in eastern Africa face droughts during cold phase of ENSO (La Niña). Dutra et al. (2013b), Lott et al. (2013), and Tierney et al. (2013) indicate that a strong La Niña event was the main cause of 2010/11 drought in the Horn of Africa. In contrast, Hastenrath et al. (2007) argue that in eastern Africa the low rainfall occurs during fast westerlies winds, which are usually accompanied by anomalously cold waters in the northwestern and warm anomalies in the southeastern extremity of the equatorial Indian Ocean Basin. This mechanism was found responsible for the 2005 drought in East Africa. Tierney et al. (2013) suggests that the Indian Ocean is an important driver for the variability of East African rainfall by altering the local Walker circulation.

Droughts in the Sahel are caused by an array of complex processes and feedback mechanisms. Caminade and Terray (2010) state that conditions that favour lower summer rainfall in the Sahel are: when the Atlantic Ocean is cool to the north of the equator and warm to the south; an increased magnitude and frequency of El Niño events; and an increased vertical thermal stability from a warming troposphere. Most of the studies on Sahelian droughts concur that the recent severe droughts in the Sahel were caused by the ocean warming (southward warming gradient of the Atlantic ocean and steady warming of the Indian Ocean) and a southward shift of Inter Tropical Convergence Zone (ITCZ) (Caminade and Terray, 2010; Dai, 2011; Giannini et al., 2008; Lebel et al., 2009; Zeng, 2003). The land-atmosphere feedbacks through natural vegetation and land cover change, and anthropogenic contribution to land use change are also important factors. Furthermore, human induced green house gas emission is also considered as a contributory factor to ocean warming (Dai, 2011).

Limited studies are available on causes of droughts in northwestern Africa. Touchan et al. (2008) suggests that neither SST nor ENSO show a clear relationship with drought patterns. Touchan et al. (2011) argues that anthropogenic green house gas emissions and related climatic change is an important factor, causing drier than normal conditions in this region. These studies suggest that the causes of droughts in northwestern Africa are currently not well established and require further research.

1.1.3 Drought management

In view of the severe drought conditions and water shortage that can be expected in the coming years in sub-Saharan Africa, efforts should focus on improving drought management. Drought management in the past has focused primarily on the response to drought events, but it is now clear that a new paradigm on drought management based on improving resilience and preparedness to drought by early warning should be adopted. Improving resilience and preparedness is imperative to reduce drought vulnerability and risk to the societies. Vulnerability can advance a natural hazard into a disaster. A complete and effective early warning system comprises four elements: knowledge of the risks involved, monitoring and forecasting, dissemination and communication, and response capability. Failure in any of these parts may result in failure of the whole early warning system (UNISDR, 2006). Of course, early warning should be supported by established and suitable institutional frameworks and drought policies. Moreover, education is vital. If society is not educated on how to respond to the forecasts and warnings, then potential damages and losses will hardly be reduced. The costs and damages related to droughts can be extremely high.

In the last decades there has been an increasing worldwide awareness on droughts. During a high-level meeting on national drought policies hosted by the World Meteorological Organization (WMO) in Geneva in March 2013 several strategies were adopted on drought management. These clearly identified forecasting and warning as key measures (HMNDP, 2013; UNCCD et al., 2012). This would require advancing all four of the elements of an effective early warning system. Operational forecasting is already commonplace in several parts of the world, but the main focus is often on floods. However, the relevance and importance of drought forecasting is gaining attention in the research community, and increasingly so in operational practice.

1.1.4 Drought classification and characterisation

Droughts are often classified into four types: meteorological, agricultural, hydrological, and socio-economic (AMS, 1997; Wilhite and Glantz, 1985; Mishra and Singh, 2010). Although the first three indicate droughts as physical phenomena, the last one associates the physical phenomena of droughts to their impacts on people and environment. A *meteorological drought* occurs when accumulated precipitation for a defined period (normally on the scale of months) is lower than that in a normal year, which is usually defined by a long-term average (Dai, 2011). *Agricultural drought* links the various characteristics of meteorological drought to agricultural impacts, focusing on precipitation shortages, differences between actual and potential evaporation and soil-water deficits that can lead to crop failure (AMS, 2004; Motha, 2011). A few days or weeks of a lack of moisture in the root zone, especially during the growing season, may already create stress on crops resulting in reduced crop yields. Agricultural droughts can also be triggered or aggravated by other meteorological conditions such as high temperatures, wind, and low relative humidity (Heim, 2002; Teuling et al., 2013).

Hydrological droughts are concerned with the effects of periods of precipitation shortfall on surface or subsurface discharges and water resources, rather than with precipitation shortfalls directly (AMS, 2004). They take place as reduced streamflows, lowered water levels in lakes and reservoirs, and lowered groundwater tables. Hydrological droughts are typically out of phase, usually lagging behind the occurrence of meteorological and agricultural droughts (AMS, 2004). They also have a much larger inertia than meteorological drought. For example, a storm may be sufficient to recover from meteorological drought but insufficient to replenish streams and lakes, and bring them to normal levels, which means the hydrological drought may persist.

Socio-economic droughts refer to the impacts of the other types of droughts (meteorological, agricultural, and hydrological) on social and economic aspects of the population affected (AMS, 2013). Therefore, socio-economic drought is a result of the physical drought conditions (as a hazard) and vulnerability of the society to drought hazards. The connection between the different types of droughts is obvious, but their relationships are rather complex. It is not straightforward how one affects the other and, more specifically, if and when one type of drought (e.g., meteorological) may lead to another type of drought (e.g., hydrological).

The definition of drought as "a prolonged period of abnormally dry weather condition leading to a severe shortage of water" (Maskey and Trambauer, 2014), presented earlier in this chapter, is very qualitative and general. More specific questions we could ask are: How long does the condition need to continue to call it "prolonged"? When do we call a weather condition "abnormal"? And when do we call the water shortage "severe"? To answer these questions, different drought indicators have been proposed in the past (Heim, 2002; Zargar et al., 2011). A drought indicator is often a standardised numerical value based on anomalies of a certain parameter representing the availability of moisture or water (e.g., precipitation, soil moisture, and streamflow) from its long-term mean (NDMC, 2015). Thus, a drought indicator is used to characterise a drought quantitatively using parameters such as its onset, termination, and intensity. A drought indicator often measures the intensity of a drought event, which together with its duration (i.e., the date of termination minus the date of onset) define the severity (or magnitude) of the event (McKee et al., 1993). Various indicators have been developed to characterise different types of droughts (e.g. Standardized Precipitation Index (SPI, McKee et al., 1993), Palmer Drought Severity Index (PDSI, Palmer, 1965), Standardized Precipitation Evaporation Index (SPEI, Vicente-Serrano et al. 2010a, 2010b), Standardized Runoff Index (SRI, Shukla and Wood, 2008)).

Many of the drought indicators are based on the outputs that can be obtained from a hydrological model (e.g. evaporation, soil moisture, and runoff). Although some observation data may be available for hydrological drought monitoring and analysis of past droughts, such data are generally limited to few locations usually insufficient to cover the spatial variability of droughts. Distributed or semi-distributed hydrological models can then be very useful in complementing the available data. Moreover, a hydrological model can be used as a predicting tool, which can assist in operational drought forecasting. Thus, there is significant potential in the use of hydrological models for drought assessment, monitoring, and forecasting.

1.1.5 Hydrological model for drought assessment

Hydrological models simulate water flux and storage through various media within the hydrological cycle. Precipitation and temperature are the major driving inputs for all hydrological models. Depending on how physical processes are represented, additional inputs such as solar radiation, relativity humidity, and wind speed may be required. The outputs from a process-based or conceptual hydrological model range from (actual) evaporation, soil moisture, and groundwater recharge through to reservoir inflow and river runoff. Moreover, the runoff can be obtained in three different components; surface runoff, interflow and baseflow. The outputs from hydrological models can be used for the computation of hydrological or agricultural droughts indicators for drought assessment. However, not all hydrological models sufficiently represent hydrological processes that are important for characterising droughts in a given climatic condition. Processes like evaporation or surface water - groundwater interactions are very important for drought assessment. As a result, a detailed process-based, distributed or semi-distributed model is preferred with a continuous simulation capability. Models that are aimed at applications for floods may not have a good evaporation component or may only be used for an event simulation (contrary to continuous simulation), and would therefore not be suitable for drought assessment. Therefore, if results of a hydrological model are to be used for drought assessment, the model should be carefully selected considering the spatial scale, data availability and end-user requirements (Trambauer et al., 2013).

Despite the fact that hydrological models have a potential to provide useful information for drought assessment, they are not always used. Drought indicators that are based on meteorological data and in some cases meteorological models are more commonly used. Meteorological drought indicators are easier to compute, and when aggregated for longer periods, have shown to be highly correlated with agricultural and hydrological droughts. However, there is little research on whether a hydrological model adds value to identification and forecasting of hydrological droughts in a region or basin, and has to be carefully explored.

1.1.6 Drought forecasting and early warning

Early warning is imperative to help mitigate socio-economic and environmental impacts of droughts. When warnings are issued several months ahead of a drought event, water managers can trigger action plans and raise preparedness to reduce the risk of severe drought impacts. To date, decisions in water management at the end user level are often based on the current state. However, in order to extend lead time and thus increase the effectiveness of mitigation measures, seasonal forecast information is currently promoted (e.g. www.DEWFORA.net) and several meteorological services already provide seasonal forecasts, e.g. ECMWF System S4 (Molteni et al., 2011) (Seibert and Trambauer, 2015).

Seasonal streamflow forecasts traditionally used statistical methods (Landman et al., 2001; Robertson and Wang, 2011; Pagano et al., 2009). More recently, meteorological seasonal forecasts from global climate models have become available operationally and these are used to drive detailed hydrological models for both deterministic and probabilistic streamflow forecasts. A probabilistic streamflow forecast normally makes use of an ensemble forecasting system. An ensemble forecasting system seeks to assess and provide useful information on the uncertainty of hydrological predictions by proposing, at each time step, an ensemble of forecasts from which one can estimate the probability distribution of the predictant, in contrast to a deterministic forecast for which no distribution is obtainable (Velázquez et al., 2011). Even though it may seem easier to make a decision from a deterministic forecast (single estimate) than from a probabilistic one, this is not necessarily the case. This is because when the forecaster issues a deterministic forecast, the underlying uncertainty is still there even though it is not presented, and then it is up to the forecaster to make a best estimate of the likely outcome. However, the forecaster's best estimate may not be well adjusted to the real need of the user (WMO, 2012b). Moreover, several studies (e.g. Cloke and Pappenberger, 2008) have shown that ensemble predictions have greater skill and capability to issue (flood) warnings than deterministic forecasts.

In the last decade several groups have dedicated significant effort into the development of probabilistic forecasts (e.g. Hydrological Ensemble Prediction EXperiment (HEPEX), www.hepex.org). Operational forecasting of streamflow is already common in several parts of the world, but the main focus is often on floods. To date this approach has not been applied as widely to drought forecasts, but the relevance and importance of drought forecasting is gaining attention. In recent years several drought monitoring and early warning systems have been established around the world, which focus on different types of drought and on different regions. Some of these are still under development and remain experimental at this stage. A short review on these existing drought early warnings systems (to the author knowledge) is presented hereafter.

The Drought Early Warning Systems (DEWS) currently in existence in the world are arguably less developed than many flood early warning systems. Grasso (2009) reports that only three institutions provide information on the occurrence of major droughts at the global scale; FAO's Global Information and Early Warning System on Food and Agriculture (GIEWS), the Humanitarian Early Warning Service (HEWS) operated by the World Food Programme (WFP), and the Benfield Hazard Research Centre at University College London.

In the United States the US Drought Monitor (http://droughtmonitor.unl.edu/) was set up in collaboration between the US Department of Agriculture (USDA), NOAA, the Climate Prediction Centre, and the University of Nebraska. It provides insight to current drought conditions and impacts at the national and state level through an interactive map, presenting multiple drought indicators combined with field information and expert input. It also includes 6–10-day outlooks and monthly and seasonal forecasts of precipitation, temperature, soil moisture and streamflow. The National Weather Service's National Center for Environmental Prediction's (NCEP) also has a (multi-model) drought monitoring system, as well as a seasonal hydrological forecasting system running at the Environmental Modeling Center (Ek et al., 2010). Additionally, the North American Multi-Model Ensemble (NMME), which became an experimental real-time system in August 2011, is mainly focused on seasonal prediction of meteorological drought (Kirtman et al., 2013).

In Europe the European Commission Joint Research Centre (JRC) has established the European Drought Observatory (EDO, http://edo.jrc.ec.europa.eu/), which includes an interactive map viewer with drought-relevant information. It includes real-time maps of different drought indicators, including the standardised precipitation index (SPI), snow and soil moisture anomaly, and vegetation productivity anomaly. These indicators are combined in an overall indicator that is used to provide warnings and alerts. A 1-week forecast of the expected soil moisture anomaly is also provided. The Beijing Climate Center (BCC) of the China Meteorological Administration (CMA) similarly monitors the development of drought across China, with maps on current drought conditions being updated daily on their website.

The FEWS Net (Famine Early Warning System Network) for eastern Africa, Afghanistan, and Central America reports on current famine conditions, including droughts, by providing monthly bulletins that are accessible on the FEWS Net web page. However, a drought forecast is not provided. Other drought warning systems over Africa include the Botswana national early warning system (EWS) for drought (Morgan, 1985) and the Regional Integrated Multi-Hazard Early Warning System for Africa and Asia (RIMES). In the latter a drought early warning system is being adapted to identify climate and water supply trends in order to detect the probability and potential severity of drought (RIMES, 2014).

Advances regarding drought early warning systems in Africa in the last few years are remarkable. There is an increasing availability of drought monitoring and forecasting tools for decision making that can provide real-time monitoring and forecasting of drought across the continent. The Land Surface Hydrology Group at Princeton University, USA, has recently established an African Flood and Drought Monitor (http://stream.princeton.edu/) with support from the International Hydrology Program of UNESCO. The system provides near-real-time monitoring of land surface hydrological conditions based on the Variable Infiltration Capacity (VIC) model. The monitor is updated every day at 2 days behind real time, and provides daily conditions of precipitation, temperature, wind speed, soil moisture, evaporation, radiation, and different components of runoff in the continent, as well as historic hydrological records in eastern, southern and western African regions for up to 10 antecedent years, and derived products such as current drought conditions. They also provide precipitation, temperature and SPI forecasts (Sheffield et al., 2014). Recently, Barbosa et al. (2013) developed a pan-African map viewer for drought within the framework of the DEWFORA project, following the main features of the earlier developed EDO. The African Drought Observatory (ADO) is a web application hosted by JRC (http://edo.jrc.ec.europa.eu/ado/ado.html) that provides historical and near-real-time monitoring information, as well as seasonal forecasts describing meteorological, agricultural and hydrological droughts (Barbosa et al., 2013).

Despite the notable advances on drought monitoring and forecasting in Africa, the use of forecasting tools in decision making is still limited, in some cases by the response capability. For instance, the 2010/11 drought in the Horn of Africa was well predicted by European Centre for Medium-Range Weather Forecasts (ECMWF), but this information was not timely used for better preparedness and mitigation of the drought, which caused a heavy toll, affecting about 12 million (Dutra et al., 2013b).

1.2 Motivation: the DEWFORA project

This research was carried out in the framework of the DEWFORA project (Improved Drought Early Warning and FORecasting to strengthen preparedness and adaptation in Africa, http://www.dewfora.net/). DEWFORA was funded by the Seventh Framework Programme for Research and Technological Development (FP7) of the European Union with a total duration of 3 years, from January 2011 until December 2013. The DEWFORA consortium was composed of 19 partners, 9 from Europe and 10 from Africa. The main aim of the DEWFORA was to reduce vulnerability and strengthen preparedness to droughts in Africa by advancing drought forecasting, early warning and mitigation practices. It mainly focused in four African basins: Limpopo, Nile, Niger, Oum er Rbia, as well as on a Pan-African scale. The DEWFORA framework addresses monitoring, predicting, timely warning and response to droughts at the seasonal time scale, applicable within the institutional context of African countries. The project contributed through improved methods for identification of vulnerable regions taking into account the increased hazard due to climate change, and feasible adaptation measures.

The DEWFORA team developed a protocol for drought forecasting and warning (see Figure 1-2) that demonstrates a strong emphasis on science, but also considers the vulnerability of society to drought and how society can benefit from drought warning, as well as how that science can be adopted into policies. The project also emphasised outreach and capacity development. For example, the drought related online courses that were developed in the project are now offered by organisations such as the UN Convention for Combating Desertification (UNCCD), and can be followed by anyone interested for free.

The DEWFORA protocol gathers information on vulnerability and pending hazard so that early warnings can be declared at sufficient lead time and drought mitigation planning can be implemented at an earlier stage. The concept is summarised in Figure 1-3 below and entails the issuing of a drought warning before the effects of drought manifest themselves.

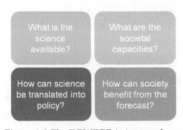

Figure 1-2 The DEWFORA Approach, an evidence-based framework for designing and implementing DEWS (Werner et al., 2015).

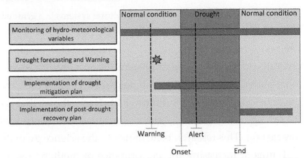

Figure 1-3 Issue of a drought warning before the effects of drought manifest (Werner et al., 2015).

The DEWFORA team evaluated drought vulnerability of agricultural systems on a pan-African scale (see Figure 1-4) in a gridded model (left map) and in a sub-basin model (right map). Both models show large similarities: high vulnerability to drought in the climatic dry north and south of the continent as well as in parts of the Rift valley and parts of west-Africa which have a less dry climate (DEWFORA, 2013a; Naumann et al., 2014). The Limpopo River basin in southern Africa is identified as being one of the most vulnerable basins of the continent.

During the DEWFORA project, drought forecasting tools were developed and in some cases operationalised. In the Niger case, for example, weekly drought bulletins are currently available. However, many mitigation and adaptation actions are still hampered by factors such as unclear institutional responsibilities, insufficient technical and scientific personnel and inadequate financial resources for adequate monitoring infrastructure and forecasting technology, and a lack of drought vulnerability analyses. Furthermore, different organisations could better coordinate their activities so that measures taken become more effective. Overall, the results from the DEWFORA project highlight that whilst early warning, mitigation and adaptation in the African region has been tackled in several ways, for it to be effective, there is a need to develop human resources involved in decision making, designing early warning systems and also to incorporate local knowledge. A discussion platform with a common language needs to be created to ensure effective communication between all stakeholders concerned (DEWFORA, 2013a).

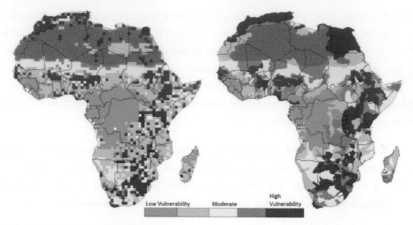

Figure 1-4 Drought vulnerability of agricultural systems in a gridded model (left map) and in a sub-basin model (right map). Areas with annual precipitation below 150 mm yr⁻¹ were masked (shaded region) (Naumann et al., 2014).

Overall, clear policies are important for an effective warning system, and the use of forecasting tools in decision making is in some cases still limited by the ability of society to adapt. However, the developed framework (Figure 1-2) clearly shows that an effective early warning system needs to be based on sound science in meteorological, hydrological and agricultural drought forecasting. This research contributes to the science available for drought early warning systems, and aims at increasing the performance of methods used for forecasting droughts in Africa by implementing state-of-the-art in (seasonal) hydrological and agricultural forecasting, and by adopting and adapting methods used in Europe, Australia and US. Moreover, given that the Limpopo River basin was identified as one of the most vulnerable basins of the continent (Figure 1-4) this research focus on the Limpopo Basin to set up the hydrological drought forecasting system.

1.3 Research objectives

The main objective of this research is to contribute to the development of a modelling framework for hydrological drought forecasting in sub-Saharan Africa as a step towards an effective early warning system.

The specific objectives of this research are to:
- Develop a framework for selecting models suitable for hydrological drought forecasting in Africa, conditional on spatial scale, data availability and end-user forecast requirements; and apply the framework to select and adapt (if necessary) a model from a wide range of existing global or continental scale hydrological models;

- Implement the selected hydrological model for the African continent and test the model performance in estimating distributed fluxes for drought assessment, in particular actual evaporation;
- Apply a higher resolution version of the hydrological model regionally for a selected basin with high drought vulnerability, and evaluate the performance of the model in identifying space-time variability of reported past droughts in the basin;
- Examine the potential of downscaling of the low resolution hydrological model results to the high resolution to assess the appropriate resolution for forecasting regional droughts; and
- Develop hydrological drought forecasting in the selected basin at a seasonal time scale by implementing a skilful probabilistic drought forecasting.

This research has both scientific and societal significance. Scientifically, it contributes to a better understanding of the tools needed for hydrological drought forecasting in the African continent for different time horizons and spatial scales. This research focuses on the science that is required to underpin an effective early warning system, in particular how hydrological drought forecasts can be developed to inform the warning process. Other aspects that are necessary in the complex process of implementing drought forecasting and warning are beyond the scope of this study (e.g. response capabilities). However, improving the science of hydrological drought forecasting contributes to effective drought warning, and contributes to societal relevant aspects such as food security, hazard management and risk reduction.

1.4 Outline of the thesis

This thesis is structured in seven chapters. In addition to the introduction and conclusions, this thesis contains five other chapters that follow the five objectives presented above. Five peer reviewed publications were developed during this study and are presented as separate chapters. These have already been published or are in the process of being published, contributing to the scientific understanding of hydrological drought forecasting.

Chapter 2 reviews several global or large-scale hydrological models and presents a decision tree for selecting the right type of hydrological model for drought forecasting in Africa.

Chapter 3 presents the results of a selected hydrological model when applied to the African continent. The resulting evaporation component, which is an important flux in hydrological drought forecasting, is compared with a number of other evaporation products.

Chapter 4 validates a high resolution version of the hydrological model for the Limpopo River basin by comparing the agricultural and hydrological droughts in the basin as simulated by the model with past observed droughts.

Chapter 5 discusses spatial scale in hydrological modelling based on the two resolution versions of the hydrological model (0.5° x 0.5° and 0.05° x 0.05°), and examines the potential of downscaling low resolution hydrological model results to a high resolution grid by considering only the signatures of the landscape at the high resolution, or by combining landscape, soil and lithological characteristics.

Chapter 6 evaluates the skill of an ensemble seasonal forecast for the Limpopo River basin, comprising the regional scale distributed hydrological model developed in Chapter 4, forced with different seasonal metrological forecasts.

2

A REVIEW OF CONTINENTAL SCALE HYDROLOGICAL MODELS AND THEIR SUITABILITY FOR DROUGHT FORECASTING IN (SUB-SAHARAN) AFRICA

This chapter provides a basis for selecting a suitable hydrological model, or combination of models, for hydrological drought forecasting in Africa at different temporal and spatial scales. Several global hydrological models are currently available with different levels of complexity and data requirements. However, some of these models may not properly represent the water balance components that are particularly important in arid and semi-arid basins in sub-Saharan Africa. This review critically looks at weaknesses and strengths in the representation of different hydrological processes and fluxes of each model. The major criteria used for assessing the suitability of the models are (i) the representation of the processes that are most relevant for simulating drought conditions; (ii) the capability of the model to be downscaled from a continental scale to a large river basin scale model; and (iii) the applicability of the model to be used operationally for drought early warning, given the data availability of the region. This review provides a framework for selecting models for hydrological drought forecasting, conditional on spatial scale, data availability and end-user forecast requirements.

This chapter is based on:

Trambauer P., Maskey S., Winsemius H., Werner M., and Uhlenbrook S.: A review of continental scale hydrological models and their suitability for drought forecasting in (sub-Saharan) Africa, Physics and Chemistry of the Earth, 66, 16-26, doi: http://dx.doi.org/10.1016/j.pce.2013.07.003, 2013.

2.1 Introduction

According to the American Meteorological Society (1997), droughts originate from a deficiency of precipitation resulting in water shortage for some activity, and its severity may be aggravated by other meteorological elements. They state that drought is a normal, recurring feature of climate, and that it occurs in virtually all climatic regimes. While aridity is a permanent feature of a regional climate, drought is a temporary aberration. Drought should be considered relative to some long-term average condition of balance between precipitation and evaporation in a particular area, a condition often perceived as "normal" (AMS, 1997; Peters, 2003).

Droughts are often grouped into four types: meteorological, agricultural, hydrological, and socio-economic (AMS, 1997; Mishra and Singh, 2010; Tallaksen and Van Lanen, 2004). Meteorological drought is defined by AMS (1997) as a lack of precipitation over a region for a period of time. They indicate that agricultural drought links the various characteristics of meteorological drought to agricultural impacts, focusing on precipitation shortages, differences between actual and potential evaporation and soil-water deficits that can lead to crop failure. Hydrological droughts are concerned with the effects of periods of precipitation shortfall on surface or subsurface discharges and water resources, rather than with precipitation shortfalls directly. Hydrological droughts are typically out of phase, lagging behind the occurrence of meteorological and agricultural droughts (AMS, 1997). They also have a much larger inertia than meteorological drought, which can basically end overnight. Socio-economic drought associates the supply and demand of some economic good with elements of meteorological, agricultural, and hydrological drought (AMS, 1997). Mishra and Singh (2010) suggest the introduction of groundwater drought as a type of drought, which has hitherto not been included in the classification of droughts. They state that a groundwater drought occurs when first groundwater recharge and later groundwater levels and groundwater discharges decrease significantly. Only major meteorological droughts will result in groundwater droughts and the lag between them can be of months or even years, much larger than the lag between meteorological and streamflow droughts (Tallaksen and Van Lanen, 2004). The spatial scale of groundwater droughts can be very variable depending on the aquifer size, recharge and discharge locations, etc. Peters (2003) concludes from her study in groundwater droughts that while short droughts will generally be more severe near the streams and dampened further away, long periods of below average recharge will have more effect near the groundwater divide.

Droughts differ in three essential characteristics - intensity, duration, and spatial coverage - and are among the most complex and least understood of all natural hazards, affecting more people than any other hazard (AMS, 1997). Africa has been severely affected in the past by intense droughts resulting in the death of hundreds of thousands of people and contributing to food insecure conditions in several African countries. In fact, a recent severe drought in 2011 (1:60 years drought) affected millions of people in the Horn of Africa. Several studies have been carried out with a view to understanding the causes of these droughts, especially in the Sahel region (Shanahan et al., 2009; Zeng, 2003; Williams and Funk, 2011; Giannini et al., 2003). Some authors claim that the intensity and severity of droughts in Africa are increasing, and attribute the cause to anthropogenic factors that lead to reduced precipitation, such as greenhouse gas and aerosols emissions (Ramanathan et al., 2001; Williams and Funk, 2011). Others claim that intervals of severe droughts lasting for decades to centuries are characteristics of the monsoon and are linked to natural variations in Atlantic temperatures (Shanahan et al., 2009). Thus, the severe droughts in recent decades are not anomalous in the context of the past three millennia, indicating that longer and more severe droughts can be expected (Shanahan et al., 2009).

For the purpose of forecasting hydrological droughts in Africa, a hydrological model should be chosen that can simulate continental hydrology, but ensuring that the hydrological processes that are important to assess droughts are considered. Various hydrological models exist at different spatial and temporal scales with diverse levels of complexity and data requirements. At the global scale a distinction can be made between Land Surface Models (LSMs) and Global Hydrological Models (GHMs). Whereas the LSMs describe the vertical exchange of heat and water, the GHMs are more focused on water resources and lateral transfer of water (Haddeland et al., 2011). By comparing simulation results of six LSMs and five GHMs in a consistent way, Haddeland et al. (2011) found that the models do not succeed in representing the water balance components in arid and semi-arid basins. Similar results were also found in other models that were not included in this comparison (Milly and Shmakin, 2002). Therefore, the selection of a suitable hydrological model, or a combination of models, for a given objective (e.g. drought forecasting in Africa) should be carried out by assessing various models using set criteria. Drought forecasting is aimed both at the continental scale and at the river basin or regional scale. Moreover, the forecasting is intended for different temporal scales: medium-range (weekly), monthly-range (1 month) and long seasonal range (up to six months). The aim of this review is to provide a framework for selecting models for drought forecasting, conditional on spatial scale, data availability and end-user forecasting requirements.

2.2   Development of the model selection framework

Five selection criteria were set for assessing the suitability of the process driven hydrological models for drought forecasting at a continental scale in Africa. The criteria are described below. These are listed in the order of importance considered in this evaluation.

1. Represented processes and fluxes

   First, the strengths and weaknesses in the representation of different hydrological processes and fluxes of the global hydrological models in question should be assessed. Ideally, a complete hydrological model would represent the following water balance components and fluxes: gross precipitation (snow, rain), interception storage, evaporation, throughfall, transpiration, snow pack storage, snowmelt, surface storage (micro depressions, lake and reservoir storage), overland flow, soil storage, recharge to shallow aquifer, capillary rise, intermediate flow, baseflow, leakage to deep aquifer, deep aquifer storage, streamflow, groundwater flow. However, there should be a compromise between model complexity and efficiency. It is important to consider the purpose of the modelling and bear in mind that a more complex model will not necessarily lead to better results. For example in drought forecasting in Africa, including the complexity of snow- water-ice dynamics will most surely not lead to better results. Representation of groundwater may, however, be of great importance.

2. Model applicability to African climatic conditions and physiographic settings

Very much linked to the previous criterion, the processes that are most relevant for simulating drought conditions in African climatic conditions and physiographic settings need to be represented. This means that processes usually considered, such as interception, evaporation, surface water-groundwater interaction and soil moisture should be included in the model while others such as glacial representation or overland fast flow are of less importance. Moreover, some extra processes or fluxes that are in general not included in the modelling framework due to its high complexity or because they are not considered important in average conditions in some regions such as channel losses, evaporation from rivers, wetlands representations, are key components for simulating droughts in specific African climatic conditions and physiographic settings.

3. Data requirements and resolution of the model (spatial and temporal resolution)

The first two criteria deal with the evaluation of the represented processes and fluxes of the models. However, including all the processes mentioned may not result in a better performance of the model if the necessary data are not available. Input data can be scarce in some regions of Africa and therefore there should be a trade-off between the data availability and process representation for drought forecasting. Some models can be very detailed and complex. However, if the necessary input data are not available then the model cannot be run in practice and extra efforts should be made to estimate the input data. This could result in a performance that is worse than that of a simpler model. For example, a detailed representation of groundwater flows and tables would be very relevant for drought forecasting in some regions of Africa, but the lack of information on the hydrogeology and groundwater tables in these regions makes this detail representation pointless, and a simpler representation is preferred.

With regards to the choice of the model grid size, there is a compromise between the need to represent spatial variability and the availability of suitable data (CEH, 2011). Most of the input data for a continental model are at a very coarse scale, and downscaling the input data to finer scales to run the model in a fine grid size leads to extra work and may not result in a higher performance. The same may be the case with the temporal resolution. Some models may have an hourly temporal resolution, but if the input meteorological data is available only at a daily resolution for example, then the use of the detailed temporal resolution may only take extra computation time without leading to better results.

4. Capability of the model to be downscaled to a river basin scale

For semi-distributed and distributed models, the grid size selection is intricately linked to the spatial scale at which the model will be applied. For example, for a continental scale model the horizontal grid resolution will more likely be in the order of several kilometres, while for a river basin the grid resolution could be in the order of one kilometre. For larger grids, processes that are only important at the local scale (such as overland flow) may not be considered in the model structure. As a result, some global models may not be easily downscaled to a river basin scale without making significant changes in the structure of the model. In the same way, it may not be possible to upscale a model that was developed for a mesoscale basin to the continental or global scale. The selected model needs to be applied at both a continental as well as at a river basin scale without important modifications in its formulation.

5. Operational model for drought early warning system at large scales

Finally, the model needs to be appropriate to use operationally, as long as the main aim of the model selection is to employ it in the development of a hydrological drought forecasting system. In this respect, a model that can easily be implemented in a forecasting environment is preferred. Hence, the model should be stable (or recover easily after failure), have reliable error and inconsistency checks, be able to run with just parts of input data (e.g. when input sources fail), be able to fit into an operational environment and should preferably be user friendly.

2.3 Assessment of available models

Hydrological models can be classified using different criteria, such as (Melone et al., 2005) (i) according to the nature of basic algorithms (empirical, conceptual or process-based), (ii) whether a stochastic or deterministic approach is taken to input or parameter specification, (iii) whether the spatial representation is lumped or (semi-) distributed, and (iv) according to the process modelled (event-driven models, continuous-process models, or models capable of simulating both short-term events and continuous simulation). In this study, a combination of conceptual and process-based (semi-) distributed hydrological models with deterministic inputs that represent continuous-process models are evaluated. The hydrological model should be suitable to evaluate the spatial and temporal occurrence of droughts based on a defined indicator. Continental and river basin-scale approaches need to be studied, and, as a result, the global macroscale model should be such that it can be downscaled to the river basin scale. Sixteen different models that are widely used or reported to be used in important applications are chosen and a brief description of each is presented in sections 2.3.1 and 2.3.2.

The macroscale models considered include five LSMs: Variable Infiltration Capacity (VIC; Liang et al., 1994), Minimal Advanced Treatments of Surface Interaction and Runoff (MATSIRO; Takata et al., 2003), Land Dynamics Model (LaD; Milly and Shmakin, 2002), ORCHIDEE (Ngo-Duc et al., 2005) and Hydrology Tiled ECMWF Scheme for Surface Exchanges over Land (HTESSEL; Balsamo et al., 2009); and eleven GHMs (or large scale hydrological models in some cases): WaterGAP (Döll et al., 2003), PCRaster Global Water Balance (PCR-GLOBWB; van Beek and Bierkens, 2009), Macro-scale-Probability-Distributed Moisture Model (Mac-PDM; Gosling and Arnell, 2010), Water Balance Model (WBM; Vörösmarty et al., 1989), Lund-Postdam-Jena model (LPJ; Gerten et al., 2004), Soil and Water Assessment tool (SWAT; Schuol and Abbaspour, 2006), SWIM (Krysanova et al., 1998), HBV (Lindström et al., 1997), Global Water Availability Assessment method (GWAVA; Meigh et al., 1999), WASMOD-M (Widén-Nilsson et al., 2007) and LISFLOOD (JRC, 2011; De Roo et al., 2000).

2.3.1 Land Surface Models

VIC is a hybrid of physically based and conceptual components, described in detail in Nijssen et al. (2001b; 2001a; 1997). The meteorological inputs for the model are daily precipitation and temperature and total evaporation consisting of three components; canopy evaporation, evaporation from bare soils, and transpiration (Liang et al., 1994). The runoff from each individual cells is combined using a routing scheme (only for the stream), to produce daily and then accumulated monthly flows at selected points. The routing model allows for the explicit representation of reservoirs. Nijssen et al. (1997) applied the model in two large basins in USA. Difficulties in reproducing observed stream flow in the arid basins were attributed to groundwater-surface water interactions which are not modelled by VIC (it does not include a mechanism to account for deep groundwater recharge and drainage to streams). The model does not have an explicit mechanism to produce infiltration excess flow and it does not represent capillary rise in the soil zone. Moreover, the processes responsible for channel losses are not represented by the routing model. This can be an important deficiency of the model in river basins like the Niger, where according to Nijssen et al. (2001a), the annually averaged flow decreases from about 1,540 m³/s to 1,140 m³/s between Kolikoro (Mali) and Gaya (Niger) even thought the catchment area at the upper point is about 10 times smaller than that at the downstream point. This difference in discharge could be also explained from high evaporation given that the Inner Niger Delta lies between these two points. The VIC model has been applied for identifying regional-scale droughts and associated severity, aerial and temporal extent under historic and projected future climate in Illinois and Indiana, USA (Mishra et al., 2010). Their results demonstrated that the major historical drought events were successfully identified and reconstructed using the model simulations. In addition, Lin (2010) reports that VIC simulated soil moisture values are used to calculate the Soil Moisture Anomaly Percentage Index (SMAPI) as an indicator for measuring the severity of agricultural and hydrological droughts. A real time drought monitoring and forecasting system for the Canadian Prairies (Lin, 2010) uses the VIC model to simulate daily soil moisture values starting from 1 January 1950 and is continually running through present with a forecast lead time up to 35 days. VIC model is also used across Africa by Princeton University to compute the African Flood and Drought Monitor.

MATSIRO has been developed for climate studies at the global and regional scales. Takata et al. (2003) present the MATSIRO model as a bucket-type hydrology model and a multilayer snow scheme. The forcing data includes wind velocity, temperature, humidity, pressure, incoming radiation and precipitation. The fluxes are calculated from the energy balance at the ground and canopy surfaces in snow-free and snow-covered portions considering a sub-grid snow distribution. The interception evaporation from canopy and the transpiration on the basis of photosynthesis are treated (Takata et al., 2003). A simplified TOPMODEL is used to calculate runoff. Four types of runoff are considered in MATSIRO: the base flow, the saturation excess runoff, the infiltration excess runoff and the overflow of the uppermost soil layer. Parameters are not considered for the groundwater part (Takata et al., 2003). The model was validated both at the global scale and at the local scale and it reproduced well the observed seasonal cycles of the energy and water balance (Takata et al., 2003). Hirabayashi et al. (2005) describe the derivation of

100-year daily estimations of terrestrial land surface water fluxes using MATSIRO. High correlations in annual runoff were obtained in most basins including the Sahel but correlations were low in dry areas and in cool-temperate zones. They believe that the poor correlations in dry areas may be due to the fact that TOPMODEL was originally developed for humid conditions. Another possible reason of low correlations in dry areas is the human effect, given that the percentage of total river water usage may be higher in dry regions (Hirabayashi et al., 2005).

LaD is a simple model of large-scale land continental water and energy balances developed by Milly and Shmakin (2002) which may be run either in stand-alone mode or coupled to an atmospheric model. Input data include incoming shortwave and long-wave radiation, total precipitation, surface pressure, and near-surface atmospheric temperature, humidity and wind speed. The model does not include precipitation interception process (Xia, 2007). Runoff is generated when root-zone soil water storage exceeds a water holding capacity. All runoff passes through a groundwater reservoir of specified residence time, and river discharge is calculated by summing all grid cells of a basin according to a river routing network (Xia, 2007). Milly and Shmakin (2002) evaluated the model and found that a few basins resulted in a major positive runoff bias that could not be explained by precipitation errors. They include, among others, the Niger River basin in the Sahel region. All of these basins are in a region where climatic aridity is strongly seasonal. The model ignores the possibility of evaporation from interception water and (except for desert) the direct evaporation from the soil. This can also lead to positive biases in runoff in arid areas (Milly and Shmakin, 2002).

ORCHIDEE solves both the energy balance and the hydrological balance (Ngo-Duc et al., 2005; d'Orgeval et al., 2008). The meteorological forcing includes precipitation, temperature, short and long-wave radiation, specific humidity, pressure and wind speed. Transpiration and interception losses are computed separately for each vegetation type, but the induced throughfall and root uptake are aggregated per vegetation class. Runoff occurs when the soil is saturated and it is the only runoff mechanism in the model. d'Orgeval et al. (2008) introduces the routing module as: surface, subsurface runoff, and river fluxes are routed through three different reservoirs in each grid box. A floodplain module is included to deal with swamps and floodplains and an optional pond module is added for small ponds that re-evaporate and re-infiltrate surface runoff over flat areas. ORCHIDEE accurately simulates most of the largest rivers. d'Orgeval et al. (2008) applied the model to an area divided in 4 regions covering different geographic characteristics (rainforest, composition of humid mountains and dry plains, semi-arid and desert) in which the sensitivity to infiltration processes was analysed. In the semi-humid basins, ORCHIDEE was found to overestimate river discharges by 20-50%, while in intermediate basins it underestimated discharge by 30-60%. In semi-arid basins ORCHIDEE was found to overestimate river discharges. Surface infiltration has a stronger impact on semi-arid regions, whereas the root zone and deep-soil infiltration resulted in having a stronger impact for semi-humid regions (d'Orgeval et al., 2008).

HTESSEL computes the land surface response to atmospheric forcing, and estimates the surface water and energy fluxes and the temporal evolution of soil temperature, moisture content and snowpack conditions. The model is forced with near surface meteorology (air temperature, wind speed, specific humidity and surface pressure) and surface fluxes (solid and liquid precipitation and downward solar and thermal radiation). Viterbo et al. (1999) point out that the surface fluxes are calculated separately for each tile, leading to a separate solution of the surface energy balance equation and the skin temperature. The latter represents the interface between the soil and the atmosphere. Below the surface, soil heat transfer follows a Fourier law of diffusion (Viterbo et al., 1999) and water movement is determined by Darcy's Law (Balsamo et al., 2009). HTESSEL is part of the integrated forecast system at the European Centre for Medium-Range Weather Forecasts (ECMWF) with operational applications ranging from the short-range to monthly and seasonal weather forecasts. A detailed description of HTESSEL can be found online and in literature (Loveland et al., 2000; Viterbo et al., 1999). The most recent version of the land surface model includes also the MODIS-Leaf Area Index monthly climatology by Boussetta et al. (2011), a bare-ground evaporation revision (Balsamo et al., 2011a) and a river routing scheme, including floodplain inundation dynamics (CaMa-Flood, Yamazaki et al. (2011)). The verification of the model's hydrology for large domains is a complex task. This is due to both the lack of direct observations and to a composite effect of shortcomings in land surface parameterizations, which produce errors not easily traced to a single process (Balsamo et al., 2009). Wipfler et al. (2011) state that the HTESSEL performs weaker in dryer areas.

2.3.2 Global Hydrological Models

WaterGAP comprises two main components: a Global Hydrology Model (including surface runoff, groundwater recharge and river discharge) and a Global Water Use Model (including withdrawal and consumptive water use; domestic, industry, irrigation and livestock) (Lehner et al., 2006). The climate input includes precipitation, temperature, number of wet days per month, cloudiness and average daily sunshine hours. A global dataset of wetlands, lakes and reservoirs was generated based on digital maps (Döll et al., 2003). Döll et al. (2003) describe the vertical water balance of the land areas by a canopy water balance and a soil water balance. In the soil water balance, capillary rise from the groundwater is not taken into account. The transport between cells is assumed to occur only as surface water flows and not as groundwater flows. In the WaterGAP Global Hydrology Model (WGHM), natural cell discharge is reduced by the consumptive water use in a grid cell as calculated by the Global Water Use Model of WaterGAP 2 (Döll et al., 2003). Only the vertical water balance for the land area is tuned by adjusting one model parameter, the runoff coefficient γ. WGHM has been tuned for 724 drainage basins worldwide, but resulting discharges were found to be overestimated in some basins, often located in arid and semi-arid areas (Döll et al., 2003). Döll et al. (2003) conclude in their study that reliable results can be obtained for basins of more than 20,000 km². However, semi-arid and arid basins are modelled less satisfactorily than humid basins. Furthermore, highly developed basins with large artificial storages, basin transfers and irrigation schemes, or basins where discharge is controlled by man-made reservoirs cannot be simulated well. Lehner et al. (2006) evaluated WaterGAP concerning its capability to assess droughts in Europe. Overall, WaterGAP demonstrated a reasonable performance in simulating timing and magnitude of average monthly and low-flow values in Europe. However, significant errors occur for certain stations and conditions.

PCR-GLOBWB is a grid-based leaky bucket type of model (coded in a dynamic modelling language that is part of the GIS PCRaster) of global terrestrial hydrology. The input data includes precipitation, actual or potential evaporation, snow and ice dynamics. Canopy interception store is finite and subject to open water evaporation. The total specific runoff of a cell consists of saturation excess surface runoff, melt water that does not infiltrate, runoff from the second soil reservoir and groundwater runoff from the lowest reservoir. The characteristic response time of the groundwater reservoir is parameterized based on a world map of lithology. River discharge is calculated by accumulating and routing specific runoff along the drainage network using the kinematic wave approximation and includes dynamic storage effects and evaporative losses from the GLWD[1] (Lehner and Döll, 2004) inventory of lakes, wetlands and plain (van Beek and Bierkens, 2009). Yossef et al. (2011) assessed the model skill to reproduce floods and droughts events in all continents and a wide range of climatic zones. They observed that the system has a markedly higher skill in forecasting floods compared to droughts, but the prospects for forecasting hydrological extremes are positive. Sperna Weiland et al. (2010) studied the usefulness of data from General Circulation Models (GCMs) for hydrological studies, with focus on discharge variability and extremes. PCR-GLOBWB was used to simulate the discharge with a GCM ensemble mean as forcing data. The resulting discharges were compared with the Global Runoff Data Center (GRDC) discharge data. Even after bias-correction, the method performed less well in arid and mountainous areas (Sperna Weiland et al., 2010).

Mac-PDM was first developed in 1999 and later further revised and improved. This model extends the well known basin-scale PDM (Probability Distributed Moisture Model) of Moore (1985). It is forced with precipitation, number of wet days (if forcing with monthly data), temperature, relative humidity or vapour pressure, net radiation and wind speed. Water that reaches the ground becomes 'quickflow' if the soil is saturated and infiltrates if the soil is unsaturated. The model assumes that all runoff generated within the grid cell reaches the cell outlet; it does not include transmission loss along the river network or evaporation of infiltrated overland flow, and does not include human intervention. The model does not route water from one grid cell to another (Gosling and Arnell, 2010). The performance of Mac-PDM.09 was evaluated by validating simulated runoff against observed runoff for 50 catchments. Because the simulated catchments are not routed, Gosling and Arnell (2010) showed that generally the runoff peak is simulated a month in advance for the larger catchments. Another result from their analysis is that the coefficient of variation (CV) of annual runoff increases with aridity, for example, the highest values are simulated over the Sahel region, amongst others. The seasonal cycle plots confirm that Mac-PDM.09 tends to overestimate runoff in very dry catchments (e.g. Niger, Murray, and Red catchments).

WBM simulates spatially and temporally varying components of the hydrological cycle and multi constituent water quality variables. Meteorological inputs to the WBM include precipitation, temperature, potential evaporation, and in more complex configurations it requires also vapour pressure, solar radiation, wind, daily minimum and maximum temperature. The WBM calculates soil moisture to a maximum defined by the field capacity of a particular soil.

---

[1] Global Lakes and Wetlands Database

When field capacity is attained, excess water is transferred to subsurface runoff pools for rain and snowmelt. From these storage pools, runoff is generated as a linear function of the existing pool size. There is no contribution to the runoff storage pools when a moisture deficit exists in relation to field capacity, and any available water recharges the soil. WBM, coupled with a water transport model (WTM) can characterize water dynamics over large areas of landscape with a high spatial and temporal resolution (Vörösmarty et al., 1989). The latest version WBMplus extends WBM by explicitly accounting for the effects of irrigation and reservoirs, implementing an improved snow melt routine, a daily time step and a Muskingum-Cunge flood routing scheme. It computes water release from large reservoirs as a function of inflow to the reservoir, mean annual inflow, current storage, and maximum capacity (University of New Hampshire, 2009). Groundwater is represented by a simple runoff retention pool that delays runoff before it enters the river channel; WBMplus does not account for the dynamics of horizontal groundwater flow or deep groundwater (Wisser et al., 2010). Fekete et al. (2004) indicate that WBM performs most poorly in extremely dry regions where rapid rain events may have the ability to produce substantial runoff despite the overall water stress.

LPJ is a dynamic global vegetation model that simulates the coupled terrestrial carbon and water cycle, and thus is well suited for investigating biosphere-hydrosphere interactions over large domains. It is driven by air temperature, precipitation, number of wet days, cloud cover, and by texture of soil types. Non-gridded model inputs include annual CO_2 concentrations. Gerten et al. (2004) compared the result of this model with three global hydrological models (WBM, Macro-PDM and WaterGAP). Their results show that the general quality of the LPJ simulation agrees well with that of the global hydrological models. Overestimations occur in semi-arid and arid regions, particularly in northern Africa, parts of South America and India. LPJ as well as the three hydrological models overestimate year-round runoff in Africa. Gerten et al. (2004) indicate that the reason for the biases in these regions are common to the GHM, where the influence of precipitation on those rivers is masked by a variety of other processes. These processes include evaporation loss (from lakes, reservoirs, wetlands, non-perennial ponds and the river channel), flood plain-channel interactions; seepage into groundwater; inter-basin transfers; and human water abstractions. These processes are not yet accounted for explicitly in global vegetation models such as LPJ (Gerten et al., 2004).

SWAT is a continuous time model in which the modelled area is divided into multiple sub-basins and hydrological response units (HRU). The meteorological forcing data includes daily precipitation, and minimum and maximum temperature. SWAT has been successfully applied for water quantity and quality issues for a wide range of scales and environmental conditions around the globe and has been shown to be suitable for large scales (Schuol et al., 2008). Schuol et al. (2008) applied the SWAT model for the whole of Africa at a monthly resolution, and calibrated and validated it at 207 discharge stations across the continent. Surface runoff is simulated using a modification of the Soil Conservation Service Curve Number (SCS-CN) Method. The runoff of each subbasin was routed through the river network to the main basin outlet. The model includes transmission losses and evaporation from the channel. Schuol et al. (2008) observed that the interannual variability of the blue water flow (surface water and groundwater) is especially large in the Sahel, in the horn of Africa and in the southern part of Africa, which are areas known for recurring severe droughts. These same areas presented also high standard deviation (SD) of the months per year without depleted green water storage (rainwater stored in the soil as soil moisture), indicating unreliable green water storage availability which often leads to reduced crop yield and consequently potential risk to frequent famines. Schuol and Abbaspour (2006) addressed some calibration and uncertainty issues using SWAT to model a four million km^2 area in west Africa. They found a large 95% prediction uncertainty (95PPU) band necessary to bracket 80% of the observed data, indicating that the uncertainty of the conceptual model is quite large. They indicated that some processes in the Niger that may be important, mainly related to the existing large reservoirs regulating the runoff of the river Niger, are not included. The large Inner Niger Delta, delaying the runoff and contributing to high evaporation losses, was also not included in the model.

SWIM is a comprehensive GIS-based tool for hydrological and water quality modelling in mesoscale watersheds (from 100 to 10,000 km^2) which was based on two previously developed tools: SWAT and MATSALU. The weather parameters necessary to drive the model are daily precipitation, air temperature and solar radiation. SWIM belongs to the intermediate class of models, combining mathematical process description with some empirical relationships (Krysanova et al., 1998). The model integrates hydrology erosion, vegetation, and nitrogen/phosphorus dynamics at the river basin scale and uses agricultural management data as external forcing. The hydrological module is based on the water balance equation and the transmission losses in the rivers are taken into account by a special module. The simulated hydrological system consists of four control volumes: the soil surface, the root zone, the shallow aquifer, and the deep aquifer. The percolation from the soil profile is assumed to recharge the shallow aquifer. Return flow from the shallow aquifer contributes to the streamflow. Krysanova et al. (1998) indicated that very flat areas with many lakes, where travel-time becomes large, are excluded in the model. Model applications (Krysanova and Wechsung, 2000) in a number of river basins in the range of about 100 to 24,000 km^2 drainage area have shown that the model is capable to describe realistically the basic hydrological processes under different environmental conditions. Krysanova et al. (1998) indicate that the model has to be further tested, especially for upscaling purposes in basins up to several thousands km^2 with 'nested' sub-basins and with different resolutions of input data.

HBV is a conceptual semi-distributed hydrological model extensively used in operational hydrological forecasting and water balance studies. It was first introduced by Bergström (1992) and later updated. The model consists of three main modules: snow accumulation and melt, soil moisture accounting and river routing and response modules (Abebe et al., 2010). The model has been applied in a wide range of scales without modification of it structure. Climatic inputs to the model are precipitation and temperature and estimates of potential evaporation. Lakes have a significant impact on runoff dynamics and the routing in major lakes is, therefore modelled explicitly. It has a simple interception storage for forested areas but interception is neglected for open areas. The response function of the model transforms excess water from the soil moisture routine to discharge to each subbasin. It consists of two reservoirs connected in series by constant maximum percolation rate and one transformation function (Lindström et al., 1997; Abebe et al., 2010). Loon et al. (2009) adapted the HBV model for the study of drought simulation in European catchments and their results shows that the HBV model reproduces observed discharges fairly well. Bergström and Graham (1998) applied the HBV model to a large scale catchment of the Baltic Sea in northern Europe, which has a total land area of 1.6 million km² with the aim of studying the possibility of upscaling the model to a continental scale and obtained successful results in calibration and validation. Love et al. (2009) indicate that even thought the HBV was developed and initially applied in Sweden for humid temperate conditions, it has also been used successfully in semi-arid and arid countries such as Australia, Iran and Zimbabwe. Furthermore, they showed the importance of interception and introduced a model structure improvement for a semi-arid basin in Zimbabwe.

GWAVA is driven by rainfall and evaporation. Model outputs include simulated monthly flows and a cell-by-cell comparison of water availability (CEH, 2011). The runoff is estimated independently for each cell and resulting flows are routed through adjacent downstream cells to derive the total flows. GWAVA has been applied to eastern and southern Africa, west Africa, the Caspian Sea basin, South America, and the Ganges-Brahmaputra basin, and is currently being applied to Europe and globally. The model incorporates additional water resource components such as reservoir operations, lakes and wetlands, groundwater abstractions, return flows, and water transfers that modify water quantity and flow regime (CEH, 2011). The routing routine includes a transmission loss term to account for reductions in river flows due to evaporation and infiltration, which can be high in semi-arid areas. Groundwater availability is assessed and water demands (population, industrial and agricultural demands) are included in the model (CEH, 2011; Meigh et al., 1999). Meigh et al. (1999) included a simple sub-model for rainfall interception losses in forested areas and an additional loss term in the groundwater component to represent drainage losses from the groundwater store. They applied the model to a region covering the whole of eastern and southern Africa, mapped water availability and demand and computed a water availability index for each country for the current conditions and for 2050 conditions. Moreover, within the PROMISE project, GWAVA was set up to model the west African region, including 22 countries and a wide range of hydrological regimes and climates. The model was set up and run to simulate baseline conditions across the region. A reasonable degree of calibration against observed flows was attained (PROMISE, 2003).

WASMOD-M model is a distributed version of the monthly catchment model WASMOD and is driven by time series of monthly precipitation, temperature, and potential evaporation. The model does not include routing delays from lakes, wetlands, and the river reach itself, as well as dam regulation (Widén-Nilsson et al., 2009). Widén-Nilsson et al. (2007) present WASMOD-M as a conceptual water-budget model with two state-variables and five tuneable parameters. Long – term-average measured runoff from 663 gauging stations in 257 basins discharging to oceans or large lakes was used for parameter-estimation and model validation. Widén-Nilsson et al. (2007) state that uncertainties and differences in model-input data, especially precipitation, are major sources of uncertainty in model output. WASMOD-M does not include regulation effects in the river basins simulation nor time-delayed routing, and therefore the inter-annual variations in basins with monthly or yearly delays are not considered.

LISFLOOD is a GIS-based hydrological rainfall-runoff-routing model. It is used in large and transnational catchments for a variety of applications, including flood forecasting, and assessing the effects of river regulation measures, land-use change and climate change. The meteorological forcing data includes rainfall, potential evaporation (for bare soil, closed canopy and open water reference surfaces), and daily mean air temperature. LISFLOOD is currently being used and tested for flood forecasting, scenario modelling, and drought forecasting (JRC, 2011). The processes simulated include: interception of rainfall by vegetation, evaporation of intercepted water, leaf drainage, snow accumulation and snowmelt, direct evaporation from the soil surface, water uptake and transpiration by plants, infiltration, preferential flow through macro-pores, surface runoff, gravity-driven vertical flow within and out of the soil, rapid and slow groundwater runoff, channel routing using kinematic (and optionally dynamic) wave. In addition, special options exist to simulate the effect of reservoirs and polders. If detailed river cross-section data are available, it is possible to use dynamic wave river routing (JRC, 2011).

2.3.3 Summary of hydrological models review

From the preceding review of hydrological models, it can be inferred that choosing an adequate macroscale tool to model the hydrology and forecast droughts in sub-Saharan Africa is not an easy task. Most of the existing global hydrological models fail to adequately represent runoff, soil moisture and other hydrological parameters in arid and semi-arid regions (Voß and Alcamo, 2008; Nijssen et al., 1997; Döll et al., 2003; Milly and Shmakin, 2002; Gosling and Arnell, 2010; Gerten et al., 2004). Several models do not represent groundwater flow and surface water-groundwater interactions including wetlands, which can be an important factor in the overall water balance of a watershed (Beckers et al., 2009). Lohmann et al. (1998) evaluated the water balances of the sixteen PILPS[2] Phase 2(c) land surface schemes (LSMs) by comparison of predicted and observed stream flow, evaporation and soil moisture changes. Their results showed that although driven by the same forcing, the models are dominated by different processes and therefore showed significant differences in their water-balance components. Responses to events were quite different as a result. However, most of the models predicted too much runoff in the dry part of the basin and, hence, under-predicted the spatial variability in the runoff fraction. In the same way, all models tended to over-predict the evaporation in winter, and under-predict it in summer. Their results suggest that most of the schemes could be improved by refining the parameterizations of soil-evaporation interactions.

In the Haddeland et al. (2011) comparison, the components of the contemporary global water balance under naturalized conditions (human impact such as reservoir and withdrawals are not included) were assessed in the simulation period 1985-1999. In their study, no major difference in the inter-annual variations were found between the models run at daily or sub-daily time steps. Major differences were also not found between models using different evaporation or runoff

---

[2] Project for Intercomparison of Land-surface Parameterization Schemes

schemes. In arid and semi-arid areas, the spread of simulated runoff and evaporation is relatively large, and the coefficient of variation is high for both evaporation and runoff. The largest absolute differences are found in the tropics, whereas the largest relative differences are found in arid areas. The resulting runoff was overestimated in the semi-arid and arid basins and they state that this may be partly due to water extractions in these areas not being considered, and to the fact that the models miss two key processes; the transmission loss along the river channel which is significant along major rivers in arid zones, and the re-infiltration and subsequent evaporation of surface runoff generated in other parts of the catchment.

Haddeland et al. (2011) show with their model intercomparison that there are considerable differences in simulated evaporation and runoff between the models, which can have a large impact on the assessment of water resources availability. They state that climate change studies need to use not only multiple climate models, but also multiple hydrological models. They conclude that when studying the impacts of climate change on the global water cycle and water resources, definite conclusions cannot be based on the results of a single model. This issue is also stressed in Hirabayashi et al. (2005).

The uncertainty in all the forcing data (mainly precipitation) is also an important issue that cannot be overlooked. Even a perfect model, if forced with biased precipitation will fail to accurately represent runoff, soil moisture and other hydrological fluxes. In Africa there are many regions with a lack of good precipitation observations, and this is a limiting factor to properly identify the limitations of each model. One way to quantify the uncertainty or error arising from input data is by using an ensemble approach (Zhu, 2005).

2.4 Discussion

With a view to providing a framework for the selection of process driven models for drought forecasting, a scheme is presented in Figure 2-1, which can be adapted for different spatial scales, climatic conditions and end user forecasting requirements. As mentioned in the selection criteria, first the process representation is critically looked at. Therefore, the decision tree starts with the list of processes that are required for an adequate simulation of drought conditions. A distinction in a second step is made with the processes that are required to simulate the hydrology in (semi-) arid regions, which is of great importance given the large coverage of these areas in the African continent. Secondly, the input data availability and possibility to use alternative data are studied. In the third step the ability of the model to be downscaled is considered. Fixed grid sizes and limitations of applicability to certain basin sizes are mainly considered here. Finally, a model that can be used operationally is preferred, meaning that it can be easily implemented in a forecasting environment. The models reviewed were assessed against the proposed criteria as just explained. A summary of this evaluation is presented in Table 2-1 and Table 2-2 for the LSMs and GHMs, respectively.

Figure 2-2 presents a stacked Venn diagram following the framework presented in Figure 2-1 for the models described and summarises the information presented in Tables 2-1 and 2-2. From this figure it can be observed that from the initial selection of sixteen macroscale hydrological models, five were selected as having higher suitability for drought forecasting in Africa. The bigger box (A) presents all the models considered in this comparison, box (B) presents the models that include the processes that are relevant for drought forecasting in Africa (or their code can be easily accessed and modified in order to include these processes). In this step, only six of the sixteen models were selected as adequate to continue in the selection process. Box (C) includes the models that were not rejected due to high-data requirements. It can be observed that in this case, Box (B) and (C) include the same amount of models, as none of them was rejected due to high-data requirements. This is due to the fact that, even though data availability is scarce in Africa, the meteorological forcing can be obtained from global data (e.g. the ECMWF reanalysis data) at a sub-daily time scale for all the climatological parameters, and also given that models like SWAT with high data requirements can also be applied in a simpler way with few parameters. The last box (D) presents the models that can be used for drought forecasting in Africa both at regional as well as at continental scale. The SWIM model is rejected here given that it is still not adequate for application at continental or global scale (but this may be modified). With regard to the last criterion as to whether the model is suited for operational purposes, all models reviewed are continuous simulation models (i.e. not event-based models), and we assume that if necessary they can be modified to be suitable for use in an operational environment. None of the models was rejected in this step.

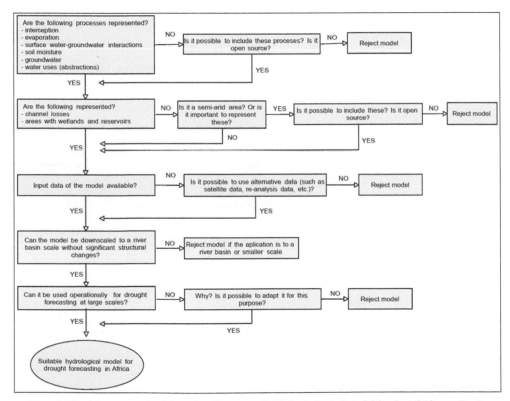

Figure 2-1 Decision tree methodology for selecting a suitable hydrological model for drought forecasting in Africa, based on the proposed criteria.

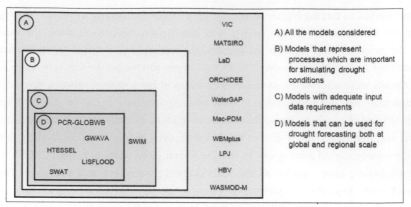

Figure 2-2 Stacked Venn diagram following the model selection procedure, starting with all the revised models in box A, and resulting with the selected models in box D.

Table 2-1 Evaluation of the five LSMs considered.

Selection criteria	LSM				
	VIC	MATSIRO	LaD	ORCHIDEE	HTESSEL
1. Represented processes and fluxes					
Represented processes					
Interception	✓ f (LAI)	✓ f (LAI)	✗	✓ f(veget cover)	✓ f (LAI)
Evaporation	✓ Penman-Monteith	✓ Bulk formula	✗✓ Energy approach	✓ Bulk formula	✓ Penman-Monteith
Snow	✓ Energy Balance	✓ Energy Balance	✓ Energy Balance	✓ Energy Balance	✓ Energy Balance
Soil	✓ 2 or 3 Layers	✓ 5 Layers	✓ 1 Layer	✓ 1 Layer	✓ 4 Layers
Groundwater	✗	✗	✓ 1 Layer	✗	✗✓ 1 Layer, linear
Runoff	✓ Satur Excess / β function	✓ Infilt and Satur Excess /GW	✓ water content excess in root zone	✓ Sat Ex	✓ VIC/ Darcy
Reservoirs, lakes	✗	✗	✗	✗	✓
Routing	✓ Linear transfer function	✗ (can be coupled ✓ to TRIP)	✗	✓ 3 fluxes routed through 3 reservoirs with 3 time constant	✗✓ Coupled with CaMa-Flood
Water use (withdraw.	✗	✗	✗	✗	✗
Energy balance	✓	✓	✓	✓	✓
Calibration parameters	✗✓ Several	✗	✗	✗	✓
2. Data requirements and resolution of the model					
Input data					
Meteorologial	Daily or sub-daily precipitation, air temperature and wind speed	6 hourly data of rainfall and snowfall rate, temperature, humidity, pressure, downward radiation and wind speed	Downward short and longwave radiation, precipitation, surface pressure, temperature, humidity and wind speed	3 hourly rainfall and snowfall rate, temperature, short and longwave radiation, specific humidity, pressure and wind speed	Rainfall and snowfall rate, temperature, short and longwave radiation, specific humidity, pressure and wind speed
Resolution of the model					
Spatial	1/16°-2° (gen 0.5°)	Varies but generally 1°	<1°	1°	>0.25° (glob 0.5°)
Temporal	Daily	Daily	Hourly and daily	30 minutes	Hourly
3. Model applicability to African conditions					
Applicability of the model in semi-arid regions	✗	✗	✗	✓	✓
4. Capability of the model to be downscaled to a river basin scale					
Model capable to be downscaled?	✓	✓	✗	✗	✓

✓ Considered
✗ Not considered
✗✓ Partially considered

Acronyms:
KWA: Kinematic wave approximation
SVE: Saint-Venant Equation
G&A inf: Green & Ampt infiltration method
VSC: Variable storage coefficient method
AW: available water

Table 2-2 Evaluation of the 11 GHMs considered.

Selection criteria		GHM										
		WaterGAP	PCR-GLOBWB	Mac-PDM	WBMplus	LPJ	SWAT	SWIM	HBV	GWAVA	WASMOD-M	LISFLOOD
1. Represented processes and fluxes												
Represented processes												
Interception		✓ f(LAI)	✓ f(veget cover)	✓ f(veget cover)	✗✓ As part of ET	✓ f(LAI)	✓ f(LAI)	✗	✗✓ (modif) HBV$_x$	✓ f(veget cover)	✗	✓ f(LAI)
Evaporation		✓ Priestley-Tailor	✓ Penman-Monteith	✓ Penman-Monteith	✓ Hamon	✓ Priestley-Tailor	✓ P-M/P-T/Hargreaves	✓ Priestley-Tailor or P-M	✓ Input	✓ Penman-Monteith	✓ From PET, AW and land	✓ Input
Snow		✓ degree day	✓ degree day	✓ degree day	✓ degree day	✓ degree day	✓ degree day	✓ degree day	✓ degree day	✓ degree day	✓ 2 temperatures thresholds	✓ degree day
Soil		✓ 1 layer	✓ 2 Layers	✓ 1 Layer	✓ 1 Layer	✓ 2 Layers	✓ ≤10 Layers	✓ ≤10 Layers	✓ 2 Layers	✓ 1 Layer	✓ 1 Layer	✓ 2 Layers 2 paralel
Groundwater		✗	✓ 1 Layer, infilt. capacity	✗	✗	✗	✓ Shallow + deep aq (GW flow eq)	✓ Shallow + deep aq (GW flow eq)	✗	✓ Monthly GW availavility	✗	✓ linear
Runoff		✓ β function	✓ Improved Arno scheme	✓ Satur Excess / β function	✓ Saturation Excess	✓ Saturation Excess	✓ modif SCS or G&A inf	✓ modif SCS	✓ Saturation Excess	✓ Sat Ex / βf	✓ f(land moisture)	✓ Infiltration Excess
Reservoirs, lakes		✓	✓	✗	✗	✗	✓	✓	✗	✗	✗	✓
Routing		✓ Constant flow velocity	✓ KWA of SVE		✓ Muskingum-Cunge		✓ VSC or Muskingum-	✓ Muskingum	✓ Muskingum-Cunge	✓ Muskingum-Cunge	✗	✓ KWA
Water use (withdrawal)		✓	✗	✗	✓	✗	✓	✗	✗	✓	✗	✗
Energy balance		✗	✓ Open waters	✗	✗	✗	✗	✗	✗	✗	✗	✗
Calibration parameter		✓ γ runoff coeff	✗	✗	✗	✓	✓ Several	✗	✓ Several	✗	✓ 5 tuneable parameters	✗
2. Data requirements and resolution of the model												
Input data	Meteorological	Monthly precipitation, temperature, no. of wet days per month, cloudiness and average daily sunshine hours	Monthly or daily precipitation, actual evapotranspiration, snow and ice dynamics	Daily or monthly precipitation, no. if wet days, temperature, relative humidity or vapour pressure, net radiation, and wind speed	Monthly precipitation, temperature and potential evapotranspiration	Monthly air temperature, precipitation, no. of wet days, cloud cover	Daily precipitation, minimum and maximum temperature	Daily precipitation, air temperature and solar radiation	Daily precipitation, temperature and estimates of potential evaporation	Monthly time series of rainfall and evaporation	Monthly time series of precipitation, temperature, and potential evaporation	Daily rainfall, potential evaporation and daily mean air temperature
Resolution of the model	Spatial	0.5°	0.5°	10 min - 2°	0.5°	In general 0.5°	Subbasins	30m and larger	Semi-distribute	0.1° or 0.5°	0.5°	100m and larger
	Temporal	Daily	Daily	Daily	Daily	Daily	Daily	Daily	Daily	Daily	Monthly	Hourly / Daily
3. *Model applicability to African conditions*												
Applicability of the model in semi-arid regions		✗	✓	✗	✗	✗	✓	✓	✗	✗✓	✗	✓
4. *Capability of the model to be downscaled to a river basin scale*												
Model capable to be downscaled?		✗	✓	✓	✓	✗	✓	✗✓	✓	✓	✗	✓

✓ Considered
✗ Not considered
✗✓ Partially considered

Acronyms:
KWA: Kinematic wave approximation
SVE: Saint-Venant Equation
G&A inf: Green & Ampt infiltration method
VSC: Variable storage coefficient method
AW: available water

2.5 Conclusions

Several hydrological models that are widely used or are reported to be used in important applications were reviewed with the purpose of assessing their suitability for drought forecasting in Africa. From the review, it can be noticed that not all of these models sufficiently represent all the important water balance components for semi-arid areas. This may be due to the fact that most models do not represent the hydrological processes that could be significant in arid regions, such as transmission losses along the river channel, re-infiltration and subsequent evaporation of surface runoff, and interception of the wet surface. Moreover, semi-arid regions are characterized with high temporal and spatial variability of rainfall, resulting in high uncertainty in rainfall estimations. This of course has an impact on the hydrological model and the simulated water balance components.

A framework for selecting models for drought forecasting was presented in this chapter and used to reduce the original selection of models to a subset of models which are considered suitable for drought forecasting, in some cases assuming some possible adaptations. The suitability of the models was assessed through applying a set of criteria that included the representation of the most relevant processes, applicability of the model to be used operationally for drought early warning with the available data, and the capability of the model to be downscaled to a smaller scale. Among the sixteen well known hydrological and land surface models selected for this review, PCR-GLOBWB, GWAVA, HTESSEL, LISFLOOD and SWAT show higher potential and suitability for hydrological drought forecasting in Africa based on the criteria used in this evaluation.

Acknowledgements

We are thankful to Florian Pappenberger and Emanuel Dutra from ECMWF for providing valuable comments to the previous version of this paper.

3

COMPARISON OF DIFFERENT EVAPORATION ESTIMATES OVER THE AFRICAN CONTINENT

This chapter applies one of the hydrological models that were found to be suitable for drought forecasting to the African continent. A continental version of the global hydrological model PCR-GLOBWB, which is based on a water balance approach, is used to compute actual evaporation for the African continent. Results are compared with other independently computed evaporation products; the evaporation results from the ECMWF reanalysis ERA-Interim and ERA-Land (both based on the energy balance approach), the MOD16 evaporation product, and the GLEAM product. Three other alternative versions of the PCR-GLOBWB hydrological model were also considered. This resulted in eight products of actual evaporation, which were compared in distinct regions of the African continent spanning different climatic regimes. Annual totals, spatial patterns and seasonality were studied and compared through visual inspection and statistical methods. The results from this chapter allow for a better understanding of the differences between products in each climatic region. Through an improved understanding of the causes of differences between these products and their uncertainty, this chapter provides information to improve the quality of evaporation products for the African continent and, consequently, leads to improved water resources assessments at regional scale.

This chapter is based on:

Trambauer P., Dutra E., Maskey S., Werner M., Pappenberger F., van Beek L. P. H., and Uhlenbrook S.: Comparison of different evaporation estimates over the African continent, Hydrol. Earth Syst. Sci., 18, 193-212, doi: 10.5194/hess-18-193-2014, 2014.

3.1 Introduction

Evaporation is one of the most important fluxes in the hydrological cycle. Recently, there has been a wide interest in estimating evaporation fluxes on a continental and global scales for a variety of purposes (van der Ent et al., 2010; Teuling et al., 2009; Miralles et al., 2011a; Vinukollu et al., 2011; Mu et al., 2011; Mueller et al., 2011; Mueller et al., 2013; Jiménez et al., 2011). The accurate estimation of these fluxes on large scales has, however, always been a difficult issue. Direct measurements of evaporation are only possible over small regions, e.g. using flux towers, and are limited to only few sites, particularly in some developed regions. FLUXNET[3] coordinates regional and global analysis of observations (CO_2, water and energy fluxes) from micrometeorological tower sites. Most of the existing global products are verified only in particular regions with available data, generally in North America and Europe (Mu et al., 2011; Alton et al., 2009; Zhang et al., 2010; Miralles et al., 2011b). Some studies have evaluated the developed global evaporation product with evaporation estimates by subtracting runoff from the precipitation (Vinukollu et al., 2011; Zhang et al., 2010). Few studies have compared the results of different evaporation products. Vinukollu et al. (2011) compared results of six evaporation products (from three process-based models forced with two radiation data sets) by computing the ensemble mean and product range globally, and by comparing the annual totals of each product over latitude bands. They found the highest uncertainties between the products in tropical and subtropical monsoon regions including the Sahel. They show that the model ensemble tends to overestimate the inferred evaporation values (inferred as P-Q). They indicate that no single model does better than any other globally, and that overall all data sets are likely to be high, which may be due to lack of soil moisture limitation in the models.

The LandFlux initiative, supported by GEWEX (http://www.gewex.org/) is clearly dedicated to evaporation. In the framework of this initiative, several global evaporation data sets were evaluated and compared (Jiménez et al., 2011; Mueller et al., 2011), and global merged benchmarking evaporation products were derived (Mueller et al., 2013). Mueller et al (2013) derived their benchmark evaporation product using 40 distinct data sets over a 17 year period (1989-2005) and 14 data sets over a seven year period (1989-1995) derived from diagnostic data sets, land-surface models, and reanalysis data. Ghilain et al. (2011) present the instantaneous (MET) and daily (DMET) evaporation products developed in the framework of EUMETSAT's Land Surface Analysis Satellite Application Facility (LSA-SAF). The MET and DMET products became operational in August 2009 and November 2010 respectively, and were satisfactorily validated against ground observations in Europe. The products were compared with models from ECMWF and from the Global Land Data Assimilation System (GLDAS) in Africa and parts of South America. This comparison showed that the spatial correlation of the products with ECMWF remained very high (85 to 95 %) and was constant throughout the whole year. However, they found that for northern and southern Africa their product (LSA-SAF MET) exhibited lower estimates than ECMWF and GLDAS, with the difference with the ECMWF product being the largest (EUMETSAT, 2011). To our knowledge, none of the existing studies regarding large-scale

---

[3] http://fluxnet.ornl.gov/

evaporation has focused on the African continent. This chapter introduces a thorough comparison of different evaporation products over diverse African regions and climates.

In most cases, estimations of actual evaporation at a continental scale rely on complementary products such as (i) remote sensing, (ii) continental-scale hydrological models, or (iii) land-surface models. However, in some ways, these three different data sources follow different theoretical basis or approaches in estimating evaporation. For example there is a significant difference in the model objective of land-surface models and hydrological models. The former focus on providing boundary conditions (turbulent fluxes) to the atmosphere (mainly focusing on the energy balance, e.g. Bastiaanssen et al., 1998a, 1998b) whereas the latter focus on closing the terrestrial water balance (Overgaard et al., 2006). In this study the class of hydrological models is represented by the PCRaster Global Water Balance model (PCR-GLOBWB, van Beek and Bierkens, 2009). This model is based on the water balance approach that focuses on water availability and vertical and lateral transfer of water. The class of land-surface models is represented by the ECMWF reanalysis ERA-Land (ERAL, Balsamo et al., 2012) and ERA-Interim (ERAI, Dee et al., 2011) using a land-surface model that describes the vertical exchanges of heat and water between the atmosphere and the land surface on a grid point scale (Balsamo et al., 2011a). The evaporation results of both are compared with the remote-sensing-based data (i.e. the MOD16 product by Mu et al. (2011; 2007), and the GLEAM product by Miralles et al. (2011b)). It is worth clarifying that PCR-GLOBWB and ERAL evaporation come from offline (or stand alone) simulations, while ERAI is a coupled land-atmosphere reanalysis product. The quality of the individual products can be influenced by different climatic regions. Therefore, in this study we differentiated the hydro-climatic regions in Africa and the comparison is carried out for each region.

The main aim of the present chapter is to compare different actual evaporation estimates for the African continent in order to gain a better understanding of the disparities between the different products within defined regions and the possible causes of these differences (e.g. resulting from the meteorological input data or from the model structure in the derivation of actual evaporation). This comparison can serve as an indirect validation of methods or tools used in operational water resources assessments. In this study we do not intend to evaluate whether one product is better than the others but to discriminate areas where good consistency can be found between the results of the selected models in contrast to regions where model results diverge. We seek to provide an uncertainty range in the expected actual evaporation values for the defined regions. The understanding of this range can be useful in, for example, water resources management when estimating the water balance.

3.2 Data and methods

3.2.1 Evaporation data sets

This section describes briefly the evaporation products used in this study (see Table 3-1 for a summary). These are all global products extracted for Africa at a daily temporal resolution, with the exception of the MOD16 product, which is a monthly product. The period chosen in this study for the evaporation comparison is 2000-2010, which is the period common to all the products.

Four of the evaporation products considered in this study are based on the PCR-GLOBWB hydrological model (see Table 3-1, these products are indicated by the PCR prefix) with differences in the input data or the addition of specific processes to assess their impact on the resulting evaporation product. Each of the products is described in detail below. The description of the first product based on PCR-GLOBWB also presents a general explanation of the PCR-GLOBWB hydrological model with the selected forcing data. We then describe the other PCR-GLOBWB-based products by emphasizing only the differences to the first product.

Table 3-1 Summary of African evaporation products used in this study.

Evaporation product	Provider	Input precipitation data	Potential evaporation – method	Spatial resolution	Temporal coverage
(1) PCR-GLOBWB	This study[*]	ERAI+GPCP	Hargreaves	0.5°	1 Jan 1979 -31 Dec 2010
(2) PCR_PM	This study[*]	ERAI+GPCP	Penman-Monteith	0.5°	1 Jan 1979 -31 Dec 2010
(3) PCR_TRMM	This study[*]	TRMM 3B42 v6	Hargreaves	0.5°	Since 1 Jan 1998
(4) PCR_Irrig	This study[*]	ERAI+GPCP	Hargreaves	0.5°	1 Jan 1979 -31 Dec 2010
(5) ERAI	ECMWF	ERAI	No PE input	~ 0.7°	1 Jan 1979 - near-real-time
(6) ERAL	ECMWF	ERAI+GPCP	No PE input	~ 0.7°	1 Jan 1979 -31 Dec 2010
(7) MOD16	Univ. of Montana	NASA's GMAO	Penman-Monteith	1 km	Since 1 Jan 2000
(8) GLEAM	VU Amsterdam	PERSIANN	Priestley and Taylor	0.25°	Since 1 Jan 1998[**]

[*] The evaporation product resulted from the PCR-GLOBWB hydrological model (van Beek and Bierkens, 2009) forced with different (varying) input data and conditions.

[**] The temporal coverage of GLEAM depends on which inputs are used to run the methodology. Here the record is restricted by the availability of PERSIANN precipitation.

3.2.1.1 PCR-GLOBWB

The PCR-GLOBWB evaporation product was calculated by means of a continental-scale version of the distributed global hydrological model PCR-GLOBWB (van Beek and Bierkens, 2009). PCR-GLOBWB is used at a global scale for a variety of purposes: seasonal prediction, quantification of the hydrological effects of climate variability and climate change, to compare changes in terrestrial water storage with observed anomalies in the Earth's gravity field and to relate demand to water availability in the context of water scarcity (see Sperna Weiland et al. (2012), van Beek et al. (2011), Wada et al. (2011), Droogers et al. (2012), Sperna Weiland et al. (2011)).

PCR-GLOBWB is a process-based model that is applied on a cell-by-cell basis (0.5°x 0.5°). PCR-GLOBWB is forced with potential evaporation, and actual evaporation is derived through simulation. Initially, the model converts potential reference evaporation E_0 into potential soil evaporation (ES_0) and potential transpiration (T_0) by introducing monthly and minimum crop factors. The crop factors are specified on a monthly basis for short and tall vegetation fractions, as well as for the open water fraction within each cell. These crop factors are calculated as a function of the leaf area index (LAI) as well as of the crop factors for bare soil and under full cover conditions (van Beek et al., 2011). Monthly climatology of LAI is estimated for each GLCC (Global Land Cover Characterization) type, using LAI values per type for dormancy and growing season from Hagemann et al. (1999). LAI is then used to compute the crop factor per vegetation type according to the FAO guidelines (Allen et al., 1998). We then updated the crop factors for irrigated areas using the global data set of Monthly Irrigated and Rainfed Crop Areas around the year 2000 (MIRCA2000) (Portmann et al., 2008; Portmann et al., 2010). Interception evaporation reduces potential transpiration, and the availability of soil moisture storage is responsible for the reductions of the potential bare soil evaporation and transpiration. The potential bare soil evaporation over the unsaturated area is only limited by the unsaturated hydraulic conductivity of the upper soil layer. For the saturated area the rate of evaporation cannot exceed the saturated hydraulic conductivity of the upper soil layer. Transpiration only takes place for the unsaturated fraction of the cells and depends on the total available moisture in the soil layers (van Beek and Bierkens, 2009).

The model is described in full detail elsewhere (van Beek et al., 2011; van Beek and Bierkens, 2009). We hereby describe the forcing data applied in the first version of the model used in this study (product 1). Three other evaporation products were derived also from this model with changes in either the forcing data or in the model structure (products 2, 3 and 4, Table 3-1).

Meteorological forcing

The model is directly forced with daily precipitation, temperature and reference potential evaporation as calculated from other meteorological variables (2 metre temperature, 2 metre dewpoint temperature, surface pressure, wind speed, and net radiation). Meteorological forcing was obtained from the ERA-Interim (ERAI) reanalysis data of the past 32 years (1979-2010). The ERAI precipitation data used in this study is available at a resolution of approximately 0.7° and was corrected with GPCP v2.1 (product of the Global Precipitation Climatology Project) to reduce the bias when compared to measured products (Balsamo et al., 2010; Szczypta et al., 2011). The GPCP v2.1 is available globally at 2.5° × 2.5° resolution with a monthly frequency, covering the period from 1979 to September 2009. It combines the precipitation information available from several sources (satellite data, rain gauge data, etc.) into a merged product (Szczypta et al., 2011; Huffman et al., 2009). From September 2009 to December 2010, the mean monthly ERAI precipitation was corrected using a mean bias coefficient based on the climatology of the bias correction coefficients used for the period 1979-2009. While this only corrects for systematic biases, this was the only option available at the time, as a new version of GPCP (version 2.2) was not available. The meteorological forcing was applied with the same spatial resolution of 0.5° as the model, using bilinear interpolation to downscale from the ERAI grid to the model grid, and is assumed to be constant over the grid cell.

Reference potential evaporation from reanalysis data

The PCR-GLOBWB model requires reference potential evaporation as a meteorological input, and this therefore needs to be estimated externally. There are several approaches to estimate potential evaporation, with diverse levels of data requirements and complexity, with different temporal scales, physically based and empirical, developed under specific regions or climates. The Hargreaves equation (input parameters: daily minimum, maximum and mean temperature and extraterrestrial radiation) was used because it has the advantage that it can be applied in data scarce regions, which is the case for several regions in Africa. Droogers and Allen (2002) compared Penman-Monteith and Hargreaves reference evaporation estimates on a global scale and found very reasonable agreement between the two methods (R^2=0.895, RMSE=0.81 mm d^{-1}). They suggest that the Hargreaves formula should be considered in regions where accurate weather data cannot be expected. The Hargreaves method requires less parameterization, with the disadvantage that it is less sensitive to climatic input data, with a possibly reduction of dynamics and accuracy. However, it leads to a notably smaller sensitivity to error in climatic inputs (Hargreaves and Allen, 2003). Moreover, Sperna Weiland et al. (2012) studied several methods to calculate daily global reference potential evaporation from Climate Forecast System Reanalysis (CFSR) data from the National Center for Atmospheric Research, for application in a hydrological model study. They compared six different methods and found a re-calibrated form of the Hargreaves equation (increasing the multiplication factor of the equation from 0.0023 to 0.0031) to outperform the other alternatives.

3.2.1.2 PCR_PM

The PCR_PM evaporation product results from forcing the PCR-GLOBWB hydrological model with Penman-Monteith reference potential evaporation. The Penman-Monteith (PM) method is one of the most widely used for the estimation of potential evaporation. Although this formula is in general highly recommended by the Food and Agriculture Organization (FAO) and is considered to be one of the most physically based methods, it is impacted by the site aridity and is reported to underestimate the potential evaporation in some regions of Africa and other arid regions (Sperna Weiland et al., 2012; Hargreaves and Allen, 2003). Nevertheless, in this study we estimate the potential evaporation with the Penman-Monteith method using input variables derived from the ERAI data. This is then used to force the PCR-GLOBWB hydrological model to assess the difference in the actual evaporation resulting from the different inputs in reference potential evaporation. This product differs from the first product (PCR-GLOBWB) only in the forcing potential evaporation data set used.

3.2.1.3 PCR_TRMM

This evaporation product (PCR_TRMM) results from the PCR-GLOBWB hydrological model simulation, but forced with the Tropical Rainfall Measuring Mission TRMM 3B42 v6 precipitation data, which has finer spatial resolution and is independent of the ERAI forecasting platform. Other meteorological inputs, including potential evaporation are as in the product 1 version of the PCR-GLOBWB model. Including this product in the comparison allows for assessing the impact that the precipitation forcing has on the resulting actual evaporation. The TRMM 3B42 v6 precipitation data was chosen from the available satellite-based rainfall estimates following the results of a recent study that validated six of these products and one reanalysis product (ERA-Interim) over four African basins (Thiemig et al., 2012), and found TRMM 3B42 together with RFE 2.0 (NOAA African Precipitation Estimation Algorithm) to be the most accurate products when compared to ground data.

3.2.1.4 PCR_Irrig

This product (PCR_Irrig) is also the result of the PCR-GLOBWB hydrological model when irrigation is included in the model structure. It was introduced in the study as some African countries such as Egypt, Morocco, Sudan and South Africa contain large irrigation areas. In South Africa, for example, there is a high density of small reservoirs for irrigation purposes (see McClain (2013)). The different components of evaporation (soil evaporation, transpiration and open water evaporation from reservoirs) are expected to increase as a result of irrigation practices, reaching potential evaporation rates under optimal irrigation practices. Moreover, a recent study by van Beek et al. (2011) suggested that the PCR-GLOBWB hydrological model might underestimate evaporation given that the default model does not explicitly consider irrigated areas. They compared the PCR-GLOBWB evaporation results with the reanalyses ERA-40 (reanalysis by ECMWF, a previous version of the ERA-Interim) evaporation and found ERA-40 evaporation to be consistently higher than the evaporation data simulated by PCR-GLOBWB. They attributed these differences to irrigation, as they indicated that ERA-40 (which includes data assimilation) accounts implicitly for irrigation by decreasing the temperature to compensate for the energy loss as latent heat. The original version of PCR-GLOBWB, on the other hand, includes irrigated areas using crop factors, but actual evaporation cannot exceed the available soil moisture as the additional contribution due to irrigation is not modelled. The difference between the two actual evaporation products represent the transpiration of the water applied (van Beek et al., 2011).

To include the influence of irrigation the original PCR-GLOBWB hydrological model was adapted. The irrigation requirement for the irrigated crop area within a cell is supplied through the storage of freshwater in the cell. For each cell where irrigation takes place, it is assumed that at least a small farm reservoir is included and that this provides sufficient storage to satisfy the demand. The irrigated area within each cell, water requirements and irrigation cropping patterns are extracted from the "Global map of irrigated areas" from Siebert et al. (2007) and FAO (1997). This modified version of the hydrological model serves to assess the impact of adding irrigation in the model structure on the actual evaporation results.

3.2.1.5 ERAI

ERA-Interim (ERAI) is a global atmospheric reanalysis produced by the European Centre for Medium-Range Weather Forecasts (ECMWF) which covers the period from January 1979 to present date with a horizontal resolution of approximately 0.7 degrees and 62 vertical levels. A comprehensive description of the ERAI product is available in Dee et al. (2011). The ERAI evaporation is the result from the coupled land-atmosphere simulations. The ERAI land component is the model TESSEL (Van den Hurk et al., 2000) that is coupled to the atmospheric model, therefore being forced (and providing fluxes to the atmosphere) with the ERAI forecasts of near-surface conditions (temperature, humidity, pressure and wind speed) and downward energy and water fluxes (precipitation, solar and thermal radiation). In ERAI, the LAI is used as fixed fields with no inter-annual variability.

3.2.1.6 ERAL

ERA-Land (ERAL, Balsamo et al 2012) is a global land-surface data set covering the period 1979-2010. ERAL is a land-surface only simulation (offline) carried out with HTESSEL (Balsamo et al., 2011a; Balsamo et al., 2011b), an updated version of TESSEL (that was used in ERAI), with meteorological forcing from ERAI and precipitation adjustments based on GPCP. HTESSEL computes the land-surface response to the near-surface atmospheric conditions forcing, and estimates the surface water and energy fluxes and the temporal evolution of soil temperature, moisture content and snowpack conditions. At the interface to the atmosphere each grid box is divided into fractions (tiles), with up to six fractions over land (bare ground, low and high vegetation, intercepted water, shaded and exposed snow). The grid box surface fluxes are calculated separately for each tile, leading to a separate solution of the surface energy balance equation and the skin temperature. The latter represents the interface between the soil and the atmosphere. Below the surface, the vertical transfer of water and energy is performed using four vertical layers to represent soil temperature and moisture. Soil heat transfer follows a Fourier law of diffusion, modified to take into account soil water freezing / melting. Water movement in the soil is determined by Darcy's Law, and surface runoff accounts for the sub-grid variability of orography (Balsamo et al., 2009). HTESSEL is part of the integrated forecast system at ECMWF with operational applications ranging from short-range to monthly and seasonal weather forecasts. The LAI in ERAL was prescribed as a mean annual climatology, as described by Boussetta et al. (2012).

ERAI and ERAL differ mainly in three aspects: (i) land-surface model - TESSEL in ERAI and HTESSEL in ERAL; (ii) coupling to the atmosphere - ERAI coupled and ERAL land-surface-only simulations (no feedback to the atmosphere forced with corrected precipitation); and (iii) data assimilation - none in ERAL while ERAI soil moisture assimilation scheme corrects soil moisture in the first three layers based on the 6-h atmospheric analysis increments of specific humidity and temperature at the lowest model level (Douville et al., 2000). While the first two points are difficult to evaluate, the impact of soil moisture data assimilation in ERAI can be evaluated by calculating the assimilation increments: i.e. the amount of soil moisture added (or removed) by the data assimilation system.

3.2.1.7 MOD16

Remote sensing provides an indirect method to estimate global actual and potential evaporation. MOD16 is described in detail by Mu et al. (2011; 2007), and computes potential and actual evaporation using MODIS (Moderate resolution Imaging Spectroradiometer) land cover, albedo, LAI, an Enhanced Vegetation Index (EVI), and a daily meteorological reanalysis data set from NASA's Global Modelling and Assimilation Office (GMAO) as inputs for regional and global evaporation mapping and monitoring (Mu et al., 2011). This method is an adaptation of a previous version of the evaporation algorithm by Mu et al. (2007), which is based on the remotely sensed evaporation model developed by Cleugh et al (2007).

Mu et al. (2011) computed potential evaporation with the Penman-Monteith method driven by GEOS-5 daily meteorological reanalysis inputs and MODIS derived vegetation data: land-surface temperature, LAI, gross primary productivity and vegetation indices were extracted from four different MODIS products. To derive actual from potential evaporation, Mu et al. (2007) include multipliers to halt plant transpiration and soil evaporation as follows: low temperatures and water stress (due to high vapour pressure deficit) limit the transpiration flow, and soil evaporation is limited by a complementary relationship hypothesis which defines land-atmospheric interactions from vapour pressure deficit and relative humidity (Mu et al., 2007).

This product has been validated against actual measurements in several regions. Mu et al. (2011) evaluated their algorithm using evaporation observations at 46 eddy covariance flux towers in the United States and Canada. In their paper they present the root mean square error (RMSE), correlation, and Taylor skill score for each flux tower, and they report that the average mean absolute bias values are 24.1% of the evaporation measurements. Kim et al. (2012) validated MOD16 global terrestrial evaporation products at 17 flux tower locations in Asia and found good agreement only at five locations (r = 0.50 to 0.76, bias = -1.42 to 1.99 mm 8d^{-1}; RMSE = 1.99 to 8.96 mm 8d^{-1}). They observed the best performance of the MOD16 evaporation algorithm at sites with forested land cover. They observed poor performance at sites with grassland cover in arid and polar climates. The MODIS derived potential and actual evaporation are available online (http://www.ntsg.umt.edu/project/mod16, data set retrieved in April 2012) with a resolution of 1 km and as with most standard MODIS Land products, it uses the Sinusoidal grid tiling system in which the tiles are 10° x 10° at the equator (USGS, 2012). We created the mosaic for each month obtaining one monthly map for the entire continent, and scaled it up to a resolution of 0.5° (~50 km) using the cubic convolution resampling as suggested by Keys (1981) and Liu et al. (2007).

3.2.1.8 GLEAM

GLEAM (Global Land Evaporation: the Amsterdam Methodology) is a method to derive global evaporation from a wide range of satellite observations that was developed by the VU University of Amsterdam (Miralles et al., 2011b). The version of the product used here has a spatial resolution of 0.25 degrees latitude-longitude, and it is forced with PERSIANN (Precipitation Estimation from Remotely Sensed Information using Artificial Neural Networks) precipitation data, soil moisture and vegetation optical depth data retrieved from the NASA-LPRM (Land Parameter Retrieval Model[4] – Owe et al., 2008), radiation fluxes from ERA-Interim, air temperature from AIRS (Atmospheric InfraRed Sounder) gap-filled with ISCCP (International Satellite Cloud Climatology Project – Rossow and Schiffer, 1999), and snow water equivalents from NSIDC (National Snow and Ice Data Center - Armstrong et al., 2007). GLEAM uses a modified Priestley and Taylor (PT) model, in combination with an evaporative stress module and a Gash analytical model of rainfall interception (Miralles et al., 2010), to combine the above-mentioned satellite-observable variables to derive evaporation. The GLEAM algorithm has been

---

[4]http://gcmd.nasa.gov/records/GCMD_GES_DISC_LPRM_AMSRE_SOILM2_V001.html
The data are derived from different satellite sensors: SSMI before mid-2002 and AMSR-E after mid-2002.

recently validated using measurements from 163 eddy covariance stations and 701 soil moisture sensors (Miralles et al., 2014).

3.2.2 Definition of regions

In this chapter we aim to compare the different sets of evaporation results in defined African regions. With this purpose, we divided the African continent in regions based on similar aridity conditions and annual precipitation cycles. First, we divided the African continent based on climatic classes. The classification of the different climates was done following the definition of the UNEP (1997) and the Global Aridity Index (Global-Aridity) data set produced by Zomer et al. (2008). This is published online in the Consultative Group for International Agriculture Research Consortium for Spatial Information (CGIAR-CSI) website (CGIAR-CSI, 2010). The Global-Aridity data set is provided for non-commercial use in standard ARC/INFO Grid format, at 30 arc seconds (~1km at equator). Zomer et al. (2008) calculated a global map of the mean Aridity Index (AI) from the 1950-2000 period at 0.5 degree spatial resolution as

$$AI = MAP / MAPE \qquad (3-1)$$

where MAP is the Mean Annual Precipitation (mm year[-1]) and MAPE is the Mean Annual Potential Evaporation (mm year[-1]). In their study, they computed the mean Aridity Index using the data available from the WorldClim Global Climate Data (Hijmans et al., 2005) as input parameters and the Hargreaves equation to model Potential Evaporation globally.

For the purpose of this study, we processed this AI global data set in GIS, trimmed it for the African continent, classified it into six classes according to the UNEP classification (1997) (see Table 3-2), and scaled it up to a grid resolution of 0.5 degrees in agreement with the hydrological model grid. For upscaling we used the area-majority technique, in which the pixel value that is common to majority of the input pixels (because each pixel has equal area) is assigned to the output pixel. Figure 3-1 presents the resulting map of climate classes for the African continent at the scaled up resolution.

Table 3-2 Generalized climate classification scheme for Global-Aridity values (UNEP, 1997).

Aridity Index Value	Climate Class
< 0.03	Hyper arid
0.03 – 0.2	Arid
0.2 – 0.5	Semi-arid
0.5 – 0.65	Dry sub-humid
> 0.65	Humid

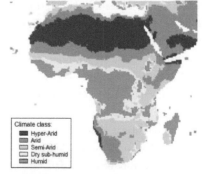

Figure 3-1 Derived climate classes for the African continent with a resolution of 0.5 degrees.

Different regions of the continent have very diverse annual precipitation cycles despite being classified in the same climate class. This is the case, for example, of the arid climate in the Horn of Africa and in south-western Africa, one characterized by two rainy seasons and another by only one rainy season in a year (see McClain (2013)). This is why most studies that divide Africa in regions usually consider sub-regions that capture the mean annual cycle of precipitation of the region (single or multiple rainy seasons, precipitation inter-annual variability, etc.). Sylla et al. (2010) divided the continent in eight regions, namely: West Sahel, East Sahel, Guinea Coast, north equatorial central Africa, Horn of Africa, south equatorial central Africa, central southern Africa, and South Africa. These regions have a uniform annual cycle of precipitation but do not distinguish between the different climatic classes within the region. The total annual precipitation in the Sahel region can be much higher for the semi-arid climate than for the arid climate (both contained in the Sahel region).

This study differentiates in regions characterized by both the regional location (e.g. Horn of Africa or southern Africa) and the climatic class within the region. We merged some of the regions defined by Sylla et al. (2010) which had very similar mean annual cycles of precipitation (e.g. East Sahel with West Sahel, central southern Africa with South Africa), to reduce the number of sub-regions from eight to five. To these we added the Mediterranean region. Six location regions were therefore defined and are presented in Figure 3-2. Finally each location region was divided in their climatic class and the resulting regions are shown in Figure 3-3. The Sahara Desert region was not considered in this study due to the negligible evaporation rates year round by virtue of hyper-arid conditions that result in very low water content in the soil.

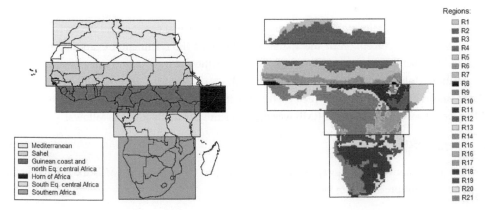

Figure 3-2 Location regions in Africa based on similarity of annual precipitation cycle (based on Sylla et al., 2010).

Figure 3-3 Derived "regions" characterized by both the regional location and the climatic class within the region.

3.2.3 Comparison of evaporation products

The common period January 2000 to December 2010 was selected to compare the evaporation products for this study. We first evaluate the difference between the crop-specific potential evaporation products that are used in this study for the derivation of the actual evaporation products (if applicable). The crop-specific potential evaporation can be defined as the amount of evaporation that would occur for a given crop if there is sufficient water available. The crop-specific potential evaporation, PET_c (mm day^{-1}) is computed within the PCR-GLOBWB model from: $PET_c = k_c \times PET_r$ where PET_r (mm day^{-1}) is the reference potential evaporation and k_c is a crop factor (dimensionless) (van Beek et al., 2011). The comparison between the actual evaporation products is then carried out at a monthly temporal scale as well as within the defined regions in the continent.

Continental maps of long-term average annual crop-specific potential evaporation were computed for each product. To compare two of the evaporation products on a gridded basis, and considering the absence of ground data to compare to, we defined the relative mean difference (RMD (%)) of the two crop-specific potential evaporation products as

$$RMD = \frac{(E_1 - E_2)}{\bar{E}} \times 100\% \tag{3-2}$$

where E_1 and E_2 is the annual crop-specific potential evaporation (mm year[-1]) data set of 1 or 2, respectively, and $\bar{E}$ is the mean crop-specific potential evaporation (mm year[-1]) of products E_1 and E_2.

RMD indicates which product is consistently higher than the other, and the relative magnitude of the difference between them (compared with the average value). This indicator seems to be a fair way of estimating the relative difference as none of the available products represents ground truth. Indicators showing absolute differences are not useful as the same absolute difference can be relatively large for areas with low actual evaporation values (arid areas), and relatively low for areas with high actual evaporation values (humid areas).

We also computed continental maps of long-term average actual evaporation for each product to allow for a visual perception of the spatial variability of the continental evaporation for each product, and between products. Moreover, to make the analysis quantitative, we created an Evaporation Multiproduct (EM) as the median of the considered products and we computed the relative mean bias (RMB (%)) between each product and the EM. In this case we considered the EM as the "observations", and the RMB was defined similarly to the RMD as follows:

$$RMB = \frac{(E_i - EM)}{EM} \times 100\% \tag{3-3}$$

where E_i is the annual actual evaporation (mm year[-1]) of data set i and EM is the annual actual evaporation multiproduct (mm year[-1]).

Subsequently, for each region the annual totals of each evaporation product were computed and the seasonality of the different products was studied and compared with the EM. The mean annual anomalies of each product with respect to the EM are presented for each region, and the variation of mean monthly actual evaporation was plotted for each product within each region and compared through visual inspection. The statistics of each evaporation product were then plotted for each region by means of Taylor diagrams of monthly evaporation and a box plot diagram of seasonal evaporation. The box plots, presented in Appendix 3-A, are displayed for the wet and the dry seasons.

3.3 Results

3.3.1 Comparison of potential evaporation products

PCR-GLOBWB model and MOD16 compute the actual evaporation fluxes from the crop-specific potential evaporation (PE) and limitations due to water availability and/or low temperatures. We hereby present three potential evaporation estimates. We computed the first two (for PCR-GLOBWB) from the ERA-Interim reanalysis data using Penman-Monteith and Hargreaves methods. The third product, MOD16 PE, uses the Penman-Monteith method but is derived from NASA's Global Modelling and Assimilation Office (GMAO) meteorological data and MODIS maps.

Figure 3-4 presents the mean annual crop-specific potential evaporation for the period 2000-2010 based on a) the Penman-Monteith method, and b) the Hargreaves method, both derived from the ERA-Interim reanalysis data. Figure 3-4 c) presents the potential evaporation from the MOD16 product. The areas in grey in the MOD16 potential evaporation correspond to missing evaporation data in the MOD16 product. MOD16 does not include urban and barren areas since there is no MODIS derived FPAR/LAI for these land cover types (Mu et al., 2011).

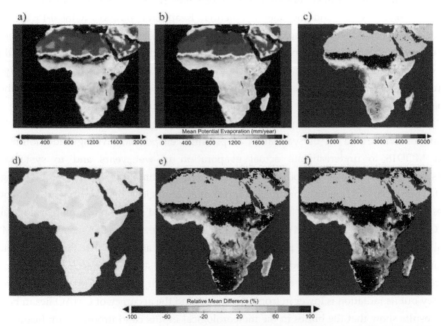

Figure 3-4 On top: comparison of mean annual crop-specific potential evaporation computations for Africa with different methods: (a) Penman-Monteith, (b) Hargreaves and (c) MOD16 product (note that the grey area corresponds to missing data and that values are presented at another scale to improve visualization). On the bottom: relative mean difference (RMD) between (d) Penman-Monteith potential evaporation (PE) and Hargreaves PE, (e) MOD16 PE and Penman-Monteith PE, and (f) MOD16 PE and Hargreaves PE.

As can be seen in Figure 3-4 a) and b), the potential evaporation derived from the Penman-Monteith equation and Hargreaves equation result in very similar values throughout the continent. The small differences are due to the different formulations of the method and the greater number of input parameters that Penman-Monteith requires, in comparison with the more simplified Hargreaves method. However, if we analyze at much smaller temporal and/or spatial scales the difference is likely to be more visible. The potential evaporation from MOD16, on the other hand, results in much higher values than the ones derived from the other two methods, especially for arid and semi-arid areas. The differences are such (a factor of 2 or up to 3) that the map needs to be presented with a different scale. The large differences between MOD16 potential evaporation product and the first product are a result of the differences in the input meteorological data (probably radiation) and vegetation data. The high disparities between the different potential evaporation products seem to be quite common, as also reported in Sperna Weiland et al. (2012), Weiß and Menzel (2008) and Kingston et al. (2009). Figure 3-4 d) to f) presents the relative mean difference (RMD) between each pair of products, and shows that the difference between the MOD16 potential evaporation product with the other two products is much smaller in humid areas than in arid and semi-arid areas. Penman-Monteith and Hargreaves products present RMD smaller than 20% throughout the continent.

We believe that the most plausible estimations for the potential evaporation could be somewhere in between, i.e. higher than Penman-Monteith and Hargreaves computed with ERA-Interim, but lower than MOD16. We also compared the Hargreaves reference potential evaporation (PEr) computed in this study and the Global Potential EvapoTranspiration (Global-PET) data set (Zomer et al. 2008), which was also computed using the Hargreaves method (selected among five different methods tested) using inputs from the WorldClim database. We observed that the PEr from the Global-PET is in general 20-30% higher than the one computed in this study. This difference should be mainly due to the difference in temperature data sets used in the two estimates. In this comparison radiation does not influence the result, because same extraterrestrial radiation values are used in both cases. Regarding the MOD16 data set, little information was found on the validation of potential evaporation, Wang and Zlotnik (2012) found MOD16 to underestimate actual evaporation in wet years and to systematically overestimate potential evaporation across Nebraska. Overestimations of MOD16 PET might be due to biases in LAI values or in the input meteorological data from GMAO, such as overestimation of solar radiation. Zhao et al. (2006) compared three known meteorological data sets: GMAO, ERA-40 from ECMWF and NCEP/NCAR to evaluate the sensitivity of MODIS global terrestrial gross and net primary production (GPP and NPP) to the uncertainties of meteorological inputs. They found that NCEP tends to overestimate surface solar radiation and underestimate both temperature and vapour pressure deficit (VPD), ECMWF has the highest accuracy but its radiation is lower in tropical regions, and the accuracy of GMAO lies in between. Their results show that the biases in the meteorological inputs can introduce significant error in the evaporation estimates.

3.3.2 Comparison of actual evaporation products

3.3.2.1 Mean annual evaporation

From the maps of mean annual evaporation for each product (not shown) it appears that the similarity between the different actual evaporation products is much higher than it is between the potential evaporation products. This suggests that the high variability introduced by the potential evaporation products is decreased in the derivation of actual evaporation. Figure 3-5 shows the maps of RMB between the mean annual actual evaporation of each product and the EM for the period 2000-2010. The RMB results in the highest values for the hyper-arid areas surrounding the Sahara Desert, region that was left out of this analysis due to its negligible actual evaporation values. On the continental scale the mean annual evaporation maps of PCR-GLOBWB and the PCR_Irrig products are almost identical. The difference is only apparent if compared on a much smaller scale. Therefore, PCR_Irrig product was not included in the EM product to avoid a double weight. ERAI is generally considerable above the EM, and for the remaining products the offset with the EM depends on the region. Over some water bodies higher evaporation values are noticeable in the products resulting from the PCR-GLOBWB model when compared to the other products; open water evaporation is considered in the total actual evaporation in this model by means of crop factors, which are specified for the fractions: open water, short vegetation and tall vegetation for each cell. ERAI and ERAL only consider water bodies bigger than 3000 km². For those grid points, only the energy balance is calculated and evaporation given as a free water surface with prescribed temperature. In GLEAM evaporation from the open water is considered as potential evaporation which is computed using Priestley and Taylor method. In MOD16 the contribution of lakes and rivers is not modelled, the evaporation therefore refers only to the land evaporation.

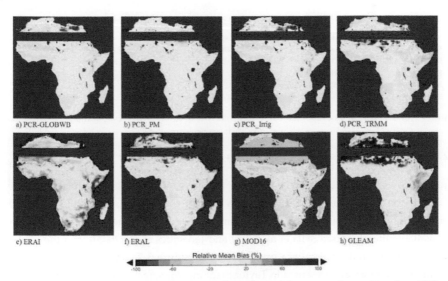

Figure 3-5 Relative mean bias (RMB) between each product and the evaporation multiproduct (EM).

The annual anomalies of evaporation for each product with respect to the EM are presented in Figure 3-6 for each region. The figure shows that ERAI evaporation product has the highest annual evaporation in the continent for almost every region with the exception of the hyper-arid and arid areas of the Mediterranean (R2 and R3) and Sahel (R4 and R5) regions, which border the Sahara Desert. Regarding the other evaporation products, in most cases they deviate from the EM for less than 100 mm year⁻¹, with some few exceptions for MOD16, PCR_TRMM and GLEAM. The PCR-GLOBWB derived products forced with ERAI+GPCP precipitation are in almost every case close to the EM with the exception of humid Sahel (R8). PCR_TRMM product (PCR-GLOBWB model forced with TRMM precipitation data) is mostly lower than the EM (with the exception of the regions bordering the Sahara Desert), and closer to GLEAM, both forced with satellite rainfall products.

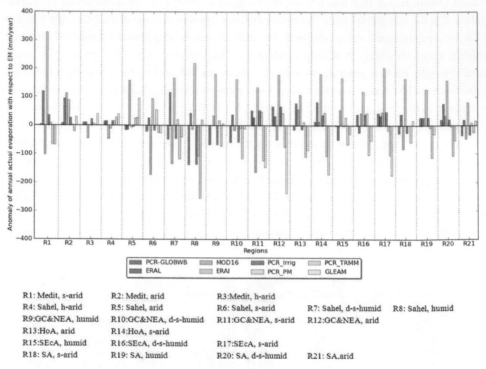

R1: Medit, s-arid | R2: Medit, arid | R3:Medit, h-arid
R4: Sahel, h-arid | R5: Sahel, arid | R6: Sahel, s-arid | R7: Sahel, d-s-humid | R8: Sahel, humid
R9:GC&NEA, humid | R10:GC&NEA, d-s-humid | R11:GC&NEA, s-arid | R12:GC&NEA, arid
R13:HoA, arid | R14:HoA, s-arid
R15:SEcA, humid | R16:SEcA, d-s-humid | R17:SEcA, s-arid
R18: SA, s-arid | R19: SA, humid | R20: SA, d-s-humid | R21: SA,arid

Figure 3-6 Annual anomaly of evaporation for each product with respect to the evaporation multiproduct (EM) for each region.

3.3.2.2 Monthly and Seasonal evaporation

Figure 3-7 shows how the different evaporation products follow the intra-annual variability. For each region the mean monthly actual evaporation for every product and the EM is plotted. Seasonality is highlighted in the figure. We defined the seasons in the continent as dry and wet (from available literature, see Sylla et al. (2010) and Jacovides et al. (2003)), and the wet seasons for each region are indicated with a grey shadow in Figure 3-7. The statistics of the monthly evaporation time series are presented in Taylor diagrams for some regions (see Figure 3-8), which summarize the ratio of the standard deviation of each product and the EM, their root mean square difference (RMSE, showed in grey curves in Figure 3-8), and the temporal correlation between each product and EM (based on the monthly values 2000-2010). The EM is considered here as the "reference" field. In the Taylor diagrams, products that are closer to the reference field have a "higher performance" than those which are farther, which in this study is interpreted as higher consistency between the products. Additionally, the statistics of the seasonal evaporation during the wet and dry seasons are presented graphically in Figure 3-10 by means of a box plot diagram, showing the variability of the seasonal evaporation for each product within each region.

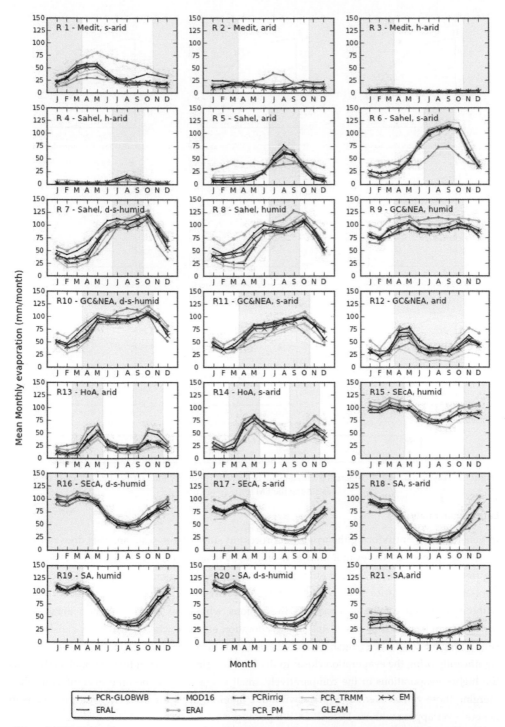

Figure 3-7 Variation of mean monthly actual evaporation for each region. In grey, the wet periods, and in white the dry periods according to Sylla et al. (2010) and Jacovides et al. (2003). MOD16 product is not presented in hyper-arid areas (R3 and R4) plots due to unavailability of data for these areas.

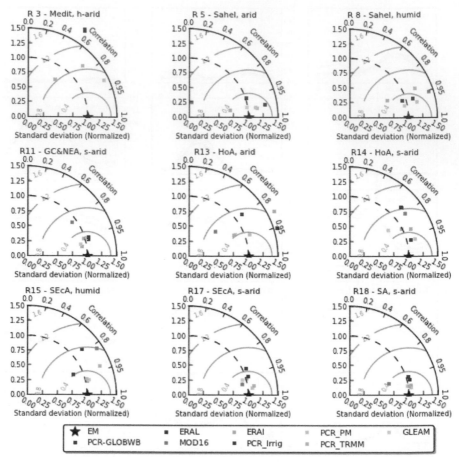

Figure 3-8 Taylor diagrams summarizing the statistics between the monthly time series of the different products assuming the EM is the "reference".

Large-scale analysis

In general terms, introducing the irrigation process in such a large-scale analysis does not result in visible changes as the evaporation is averaged over large regions. This can be observed in Figure 3-6 and Figure 3-7, where only in the Mediterranean region (R1, R2 and R3) and in the semi-arid Horn of Africa (R14) a slight deviation between the evaporation products with and without irrigation can be observed. In the regions where the larger irrigation areas in the continent are located (e.g. Nile Delta), the evaporation estimate when irrigation is considered is slightly higher than when irrigation is not considered. Introducing irrigation therefore does not significantly bring the evaporation closer to the ERA-Interim product in this regional analysis, as the higher evaporations in the comparatively small irrigated areas become insignificant when merging these values over larger areas without irrigation. In Figure 3-7 PCR_Irrig product is in general overlapping with the PCR-GLOBWB product.

Another feature that is clearly visible from Figure 3-6 and Figure 3-7 is that for almost every region and season the ERAI product consistently has the highest evaporation. ERAL evaporation is in almost every case lower than that of ERAI. Figure 3-9 shows the mean annual soil moisture increments in ERAI. The increments are positive in most of the African continent, and partly explain the higher values of evaporation in ERAI when compared with ERAL. Drusch et. al. (2008) provide a detailed evaluation of the soil moisture analysis scheme used in ERAI pointing to some of the limitations (e.g. root zone soil moisture acts as a "sink" variable, in which errors are allowed to accumulate). They also present a new surface analysis scheme that is currently operational at ECMWF. The results in this study do not allow to clearly identify the main source of the differences between ERAI and ERAL, but qualitatively, ERAL is closer to the remaining data sets.

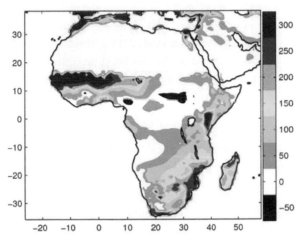

Figure 3-9 Mean annual soil moisture increments (in mm year[1]) in ERAI resulting from the soil moisture assimilation scheme.

Likewise, higher evaporation values of GLEAM bordering the Sahara Desert are due to an error in the data assimilation of surface soil moisture in deserts that has been corrected in recent versions of the product (Diego Miralles, personal communication).

Regarding the impact that input meteorological data has on the resulting evaporation, we should look at both input precipitation and input potential evaporation. With respect to the input precipitation, there seems to be a general behaviour showing that the evaporation resulting from the model forced with TRMM precipitation is consistently lower than the evaporation that results when the model is forced with ERAI+GPCP (see Figure 3-6 and Figure 3-7). For almost every region, TRMM forced model results in lower evaporations than the EM. Regarding the potential evaporation input, it can be seen that the evaporation generated with Penman-Monteith potential evaporation is in every case very similar to the one forced with Hargreaves, which was expected due to the small differences in the forcing potential evaporation products. The MOD16 evaporation show poor consistency with the other products in arid areas as also observed by Kim et al. (2012); this can be seen in Figure 3-7 (R2, R5, and R21) and Figure 3-8 (R5). For other regions, mainly in the dry sub-humid and semi-arid climatic regions, MOD16 is more consistent with the other products (see Figure 3-8).

A detailed comparison description for each region is presented in Appendix 3-A. We hereby present a general interpretation of the impact of the input data and modelling parameterizations. A full in-depth study of these including the impact of parameter values is considered to be outside the scope of this paper.

A.   Effect of input meteorological data in the estimation of actual evaporation

i.   Precipitation data

The intensity of the forcing precipitation has a large influence in the simulated actual evaporation. ERAI precipitation corrected with GPCP in general compares well with observed data at a monthly basis and has good ability to estimate the peak locations, but it is known to overestimate the frequency of wet days and underestimate the daily highest intensities in some regions (e.g. Iberian Peninsula, (Belo-Pereira et al., 2011)). Thiemig et al. (2012) obtained similar results for a number of African river basins for the ERAI precipitation data without correction. For the scale considered in this study, it is clear that rainfall events with higher intensities result in lower evaporation values (see Figure 3-6 and Figure 3-7) given that PCR_TRMM evaporation is generally lower than PCR-GLOBWB evaporation. This can be explained as higher intensities lead to higher surface runoff, which keeps the water out of reach of evaporation resulting in lower evaporation rates. Moreover, for vegetated areas, less intense rainfall tends to increase the direct evaporation as the rain is more easily intercepted by the vegetation, and thus to reduce the infiltration.

The difference between the evaporation products PCR_TRMM and PCR-GLOBWB is not negligible and it varies from region to region, and therefore is important to force the hydrological model with the most suitable precipitation product for that particular region. Thiemig et al. (2012) found TRMM together with RFE to be the best satellite products available for the African continent. However, they noted that both TRMM and ERAI underestimate the amount of rainfall during the heavy rainfall events, and they explained that for the satellite products this was mainly the result of the small extent of the heavy precipitation cells, which are generally lower than the detection limit of the satellite sensors. While ERAI highly overestimates the number of rainy days, TRMM also has some overestimation of the rainy days for tropical wet and dry zones but lower than that of ERAI (Thiemig et al., 2012). Moreover, Wang et al. (2009) highlights some known "anomalies" in TRMM such as underestimation in "warm-rain" regimes. In these regimes rain is derived from non-icephase processes in clouds (see Lau and Wu (2003) for detailed explanation).

Differences between GLEAM and the other products are partly due to the differences in input precipitation data. GLEAM is forced with PERSIANN which differs with ERA-Interim and TRMM 3B42v6 across Africa (Thiemig et al. (2012)). MOD16 is forced with GMAO precipitation, and biases in this compared to with ERA-Interim, TRMM, PERSIANN are also a source of the disparities between the evaporation products. The impact of the distinct precipitation inputs is difficult to quantify here as the products differ in other meteorological inputs (such as radiation) and they are based on other modelling approaches.

ii.   Potential evaporation data

The potential evaporation used as a forcing appears to have some influence in the estimation of actual evaporation in some of the regions, but the difference is considerably lower than the differences between the input potential evaporation products. Already quite small differences between Hargreaves and Penman-Monteith potential evaporation resulted in appreciably smaller differences in the actual evaporation products. Moreover, large differences in the potential evaporation between MOD16 and the potential evaporation inputs for PCR-GLOBWB (Hargreaves and Penman-Monteith) resulted in substantially smaller differences in the actual evaporation products (see Figure 3-7).

Biases in the input potential evaporation data are mainly due to differences in radiation, temperature, and vegetation data. Positive biases in the radiation from GMAO could be an important source for the higher values in the MOD16 potential evaporation compared to the other products. All the other products use the ERA-Interim net radiation or extraterrestrial radiation with the Hargreaves method. Regarding *temperature* inputs as stated in section 3.3.1 we observed that differences in temperature data sets between ERAI and WorldClim might be responsible for the 20-30% differences in the Global-PET data set from Zomer et al. (2008) and the one computed within this study.

B.   Effect of the model structure in the estimation of actual evaporation

The comparison between the PCR-GLOBWB and the PCR_Irrig evaporation products can help us identify the effect of introducing an irrigation scheme in a hydrological model on simulated actual evaporation. The only difference between these two models is the introduction of the irrigation scheme in PCR_Irrig which was not included in the default PCR-GLOBWB. A recent study by van Beek et al. (2011) attributed partly the difference between PCR-GLOBWB simulated actual evaporation and ERA-40 reanalysis evaporation to the under representation of the irrigation areas in PCR-GLOBWB.

It appears that introducing an irrigation scheme in the hydrological model has negligible effect in the actual evaporation results for the resolution of 0.5° x 0.5° used in this study. Only in some regions with very large irrigation areas, marginally higher evaporation was observed. This is because the higher evaporations in the comparatively small irrigated areas become insignificant when averaged over large regions containing large areas without irrigation. In a cell by cell area, the difference in evaporation when irrigation is included is noticeable. Moreover, we observed considerable differences in the evaporation results when the same irrigation scheme was introduced in a finer resolution model (0.05° x 0.05°) for the Limpopo River basin.

Regarding ERAI and ERAL, it is interesting to notice a large impact of the differences in the two model structures on evaporation estimates. As described in section 3.2.1.6, ERAI has an improved land-surface model, feedback with the atmosphere (through direct coupling) and soil moisture assimilation. In ERAI the effect of data assimilation is mostly to add soil moisture to the root zone, leading to increased evaporation.

Interception plays an important role in evaporation, and this may explain the generally lower values of MOD16 evaporation than the other products in (semi-)arid areas during the wet season. By analysing intra-annual variability we observed that the difference between MOD16 potential evaporation and the other products is highest during the dry season. MOD16 actual evaporation is generally higher than the EM during the dry season and lower during the wet season even though the MOD16 potential evaporation is higher in both seasons. This can be due to the representation of canopy intercepted evaporation. In MOD16, evaporation from the canopy is restricted by relative humidity, if the relative humidity is less than 70% no evaporation is considered from interception (Mu et al., 2011).

3.4 Discussion and conclusions

Possibilities to validate a continental evaporation product for Africa is still limited due to the inexistence of a continental-scale evaporation product based on ground measured data. In recent years there has been an increase in the amount of studies that focus on global evaporation estimates. Several new estimates were developed and validated with flux towers where available, mostly in North America and Europe, and received some indirect validation (e.g. comparison with another product) in other regions of the world. Moreover, some recent studies compare several of these estimates at a global scale, largely coming from land-surface models. The main contribution of this paper is to present an evaporation analysis focused on the African continent which serves as an indirect validation of methods or tools used in operational water resources assessments. Our analysis discriminates areas where there is a good consistency between different evaporation products and areas where they diverge. It also provides a range of variance in actual evaporation that can be expected in a given region, which may be useful in for example water resources management when estimating the water balance. Africa strongly relies on agriculture and several regions are often hit by severe droughts; evaporation estimates are key for assisting water managers in the estimation of water needed for irrigation. This paper compares different evaporation products for Africa and presents an Actual Evaporation Multiproduct at a 0.5° resolution. This EM that integrates satellite based products, evaporation results from land-surface models and from hydrological models forced with different precipitation and potential evaporation data sets, may serve as a reference data set (benchmark).

In general ERAI and MOD16 do not show good agreement with other products in most part of Africa, while the rest of the products are more consistent. ERAL is generally quite close to the EM, and the higher values of evaporation in ERAI when compared with ERAL are partly explained by the analysis of soil moisture data assimilation in ERAI. It also appears that in some regions like in southern Africa the agreement between the products is very good, which means that use of any of these products may be equally good. In other regions, such as in humid Sahel or the Mediterranean the choice of a particular product needs to be further studied as there is a larger difference between the products. These results are in agreement with the study of

Vinukollu et al. (2011) who found that the evaporation products they compared are most uncertain in tropical and subtropical monsoon regions including the Sahel.

Products compared at a monthly timescale certainly result in better outcomes than when the products are compared at a daily timescale. This study focused on the monthly and seasonal comparison, with daily comparisons considered to be beyond the scope of this study. However, monthly standard deviations of daily products differ from one product to the other and from one region to the other. A comparison of the monthly standard deviation of daily products (with the exception of MOD16 that did not have daily estimates) consistently showed that in arid and hyper-arid areas (R2, R3, R4, R5, R12, R13, and R21) ERAL shows the highest standard deviation, generally followed by ERAI and GLEAM. In these regions the mean values and variability of the standard deviation of PCR-GLOBWB derived products seem to be lower. In other regions, the standard deviations of all the products have roughly the same variability and mean values. Among the four PCR-GLOBWB derived products, the one forced with Penman-Monteith (PCR_PM) showed slightly higher values of standard deviation than the other three products. For every product and every region, a seasonality of the standard deviation can be observed, with the highest standard deviations during the wet seasons.

A potential action to improve this comparison study and the EM is to validate the products in different African regions with ground data, where available. Moreover, other available products could be added to the comparison and to the EM calculation to have more information on the variance between the products and a more consistent EM estimate. It is also recommended to compare the computed EM and the variability of the products with the global benchmark recently developed by Mueller et al. (2013). Similarly, in a basin-wide scale, long-term estimates of evaporation could be obtained from the water balance with an uncertainty estimate (Dingman, 1994).

Appendix 3-A

Regional analysis

1)   Mediterranean region (R1 to R3)

This region is characterized by higher evaporation rates in the months of March to May, after the end of the rainy season (see Figure 3-7). The evaporation peak is clear in the semi-arid region (R1), but becomes less noticeable in the arid region (R2) as the evaporation rates become lower and almost disappear in the hyper-arid areas (R3) where evaporation rates throughout the year are negligible. In the rainy season Figure 3-7 shows a clear offset between ERAI and ERAL with respect to the other products, where the first two present considerably higher evaporation rates. In the dry season, while ERAL comes closer to the other products, the offset of ERAI is still evident. This offset of ERAI in the dry periods decreases with increasing aridity, in contrast with the MOD16 product, which shows higher evaporation rates than the remaining data sets for arid areas (R2) in the dry period. In the Mediterranean region only the dry season of the semi-arid region (R1) presents quite a high variability between the different seasons for every product (see Figure 3-10). In the Mediterranean region the consistency between the products decreases as aridity increases, and in hyper-arid Mediterranean region (R3) all the products show very little consistency (see Figure 3-7).

2)   Sahel (R4 to R8)

The Sahel region is characterized by an annual evaporation cycle with one peak during the rainy season, namely from July to September (Sylla et al., 2010) (see Figure 3-7). The evaporation rates become higher and the peaks become clearer with increasing humidity. Figure 3-7 shows that only the MOD16 product in the arid region (R5) and to a lesser extent in the semi-arid region (R6) do not capture the annual evaporation cycle, presenting a relatively uniform evaporation throughout the year. This can also be seen in Figure 3-8 (R5) where the very low values of normalized standard deviation in MOD16 indicate that the amplitude of the annual cycle is underestimated. García et al. (2012) also found that MOD16 evapotranspiration product failed to capture the dynamics of evaporation in the Sahelian savannah. Figure 3-7 shows that the lowest evaporation values are observed for PCR_TRMM during the dry season but it is not clear for the wet season. Regarding the highest evaporation values, ERAI (and secondly ERAL) present these during the dry season, but a clear behaviour cannot be observed for the wet season. The semi-arid (R6), dry sub-humid (R7) and humid Sahel (R8) are the regions that present the largest variability in the inter-annual mean cycles for each product and the higher dispersion between the products in the dry seasons. Similarly to the Mediterranean region, the hyper-arid region (R4) in the Sahel shows the least consistency between all the products. For the other sub-regions all the products have a higher consistency, (see for example Figure 3-8 (R8)). Figure 3-8 shows that in the Sahel region ERAI product has a lower amplitude of the annual cycle than the EM, whereas PCR_PM and PCR_TRMM present a larger amplitude of the annual cycle than the EM. All the products however are well in phase (high correlation).

3)   Guinean coast and north equatorial central Africa (R9 to R12)

This region is characterized by a bimodal precipitation cycle with unpronounced peaks and precipitation minima. This precipitation pattern results in different evaporation cycles depending on the aridity of the region (see Figure 3-7). For the humid region (R9) all products show more or less uniform evaporation over the year; however, respective values between the products are different. In regions R10 and R11, the majority of the products follow roughly the same pattern, with slightly noticeable peaks around May and October. MOD16 deviates from this pattern, represents the amplitude fairly well in the annual cycle but remains out of phase (see Figure 3-8). Figure 3-7 shows that the peaks of the evaporation cycle become more pronounced when aridity increases. The arid area (R12) has a clear bimodal evaporation cycle. For this region (R12) the GLEAM product presents the lowest values throughout the year. During the dry season ERAI present the highest values; while during the wet season a clear behaviour cannot be observed. Figure 3-8 suggests that most of the products have a good consistency in R11 with the exception of MOD16, while in R12 the products show less consistency.

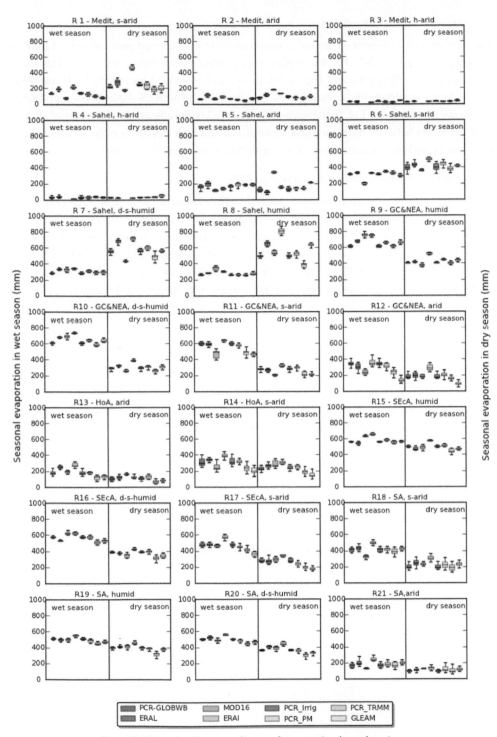

Figure 3-10 Box plot diagrams of seasonal evaporation for each region.

4)   Horn of Africa (R13 and R14)

The Horn of Africa presents a bimodal evaporation cycle with defined peaks at around May and November (see Figure 3-7). For this region most products show little consistency. This can be observed both in Figure 3-7 and Figure 3-8. In the arid Horn of Africa (R13) ERAI and ERAL present larger amplitude of the annual cycle than the EM, and MOD16 shows lower amplitude of the annual cycle than the EM. In the semi-arid region (R14), the amplitude of the annual cycle are now consistent between most products (normalized standard deviation close to one), but the correlation coefficients are rather low (the time series are not correctly phased).

5)   Southern equatorial central Africa (R15 to R17)

This region presents an evaporation seasonal cycle with a minimum in June through October and a maximum in December through April (see Figure 3-7). The difference between the evaporation values in the wet and in the dry season increases with aridity. The consistency between all products seems to be quite better in the dry sub-humid region (R16) than in the humid region (R15). In R15 MOD16 and ERAL seem to be fairly out of phase (low correlation values), and PCR_TRMM presents a larger amplitude of the annual cycle than the EM. Regarding the semi-arid region (R17), the absolute amplitude of the annual cycle increases, and quite a good consistency is observed between all the products (see Figure 3-8).

6)   Southern Africa (R18 to R21)

The southern Africa region show a pronounced seasonal cycle with a minimum in June through October and a maximum in December through April (see Figure 3-7). In the semi-arid (R18), humid (R19) and dry sub-humid (R20) regions all the products present a good consistency. In the semi-arid (R18) and arid (R21) regions MOD16 presents a lower annual cycle amplitude that the EM, which is represented in Figure 3-8 (R18) by low normalized standard deviations. In the arid region (R21) all the products show less consistency.

Acknowledgements

Special acknowledgment goes to V. Thiemig from JRC for providing pre-processed TRMM data for the African continent and to D. Miralles from the University of Bristol for providing the global GLEAM data set. The authors would like to thank G. Balsamo for the suggestions that helped to improve the manuscript.

4

IDENTIFICATION AND SIMULATION OF SPACE-TIME VARIABILITY OF PAST HYDROLOGICAL DROUGHT EVENTS IN THE LIMPOPO RIVER BASIN

In this chapter a finer-resolution version (0.05° x 0.05°) of the continental-scale hydrological model PCR-GLOBWB was set up for the Limpopo River basin, which is one of the most water stressed basins on the African continent. The finer resolution model was used to analyse hydrological droughts in the Limpopo River basin in the period 1979-2010 with a view to identifying severe droughts that have occurred in the basin. Two agricultural drought indicators and two hydrological drought indicators were computed. Other more widely used meteorological drought indicators were also computed. The indicators considered are able to represent the most severe droughts in the basin and to some extent identify the spatial variability of droughts. Moreover, results show the importance of computing indicators that can be related to hydrological droughts, and how these add value to the identification of hydrological droughts and the temporal evolution of events that would otherwise not have been apparent when considering only meteorological indicators.

This chapter is based on:

Trambauer P., Maskey S., Werner M., Pappenberger F., van Beek L. P. H., and Uhlenbrook S.: Identification and simulation of space-time variability of past hydrological drought events in the Limpopo river basin, southern Africa, Hydrol. Earth Syst. Sci., 18, 2925-2942, doi: 10.5194/hess-18-2925-2014, 2014.

4.1 Introduction

Droughts are a widespread natural hazard worldwide, and the societal impact is tremendous (Alston and Kent, 2004; Glantz, 1987). Recent studies show that the frequency and severity of droughts seems to be increasing in some areas as a result of climate variability and climate change (IPCC, 2007b; Patz et al., 2005; Sheffield and Wood, 2008; Lehner et al., 2006). Moreover, and probably more importantly, the rapid increase in world population will certainly aggravate water shortage on local and regional scales. The study of droughts and drought management planning has received increasing attention in recent years as a consequence.

Drought monitoring is a key step in drought management, requiring appropriate indicators to be defined by which different types of drought can be identified. Meteorological, agricultural and hydrological drought indicators are available to characterise different types of droughts. The best known indicators are the Standardized Precipitation Index (SPI; McKee et al., 1993) and the Palmer Drought Severity Index (PDSI; Palmer, 1965; Dube and Sekhwela, 2007; Alley, 1984); both are primarily meteorological drought indices. The SPI uses only precipitation in its computation, and the PDSI uses precipitation, soil moisture and temperature. However, the timescale of drought that the PDSI addresses is often not clear (Keyantash and Dracup, 2002), and will usually be determined by the timescale of the data set; Vicente-Serrano et al. (2010b) indicate that the monthly PDSI is generally correlated with the Standardized Precipitation Evaporation Index (SPEI) at timescales of about 9-12 months. While the computation of the PDSI is complex, applied to a fixed time window and difficult to interpret, the SPI is easy to compute and to interpret in a probabilistic sense, is spatially invariant and can be tailored to a time window appropriate to a user's interest (Guttman, 1998). Alley (1984) and Vicente-Serrano et al. (2010b) also highlight several limitations of the PDSI, such as not allowing for the distinction of different types of drought (i.e., hydrological, meteorological and agricultural) as it has a fixed temporal scale. The PDSI has other derivatives such as the Palmer Hydrological Drought Index (PHDI) for hydrological long-term droughts, Palmer Z Index for short-term monthly agricultural droughts, and the Crop Moisture Index (CMI) for short-term weekly agricultural droughts. The empirical PDSI method developed in the United States, is still widely used in the United States but is gradually being substituted by other indicators in other regions (Keyantash and Dracup, 2002) as a result of its limitations. The SPI can be computed for different timescales by accumulating the precipitation time series over the time period of interest (typically 3 months for the SPI-3, 6 months for the SPI-6, and 12 months for the SPI-12). SPI has shown to be highly correlated with indicators of agricultural drought, hydrological drought and groundwater drought. The SPI-3 has a high temporal variability that is associated with short-to-medium range meteorological anomalies that can result in anomalous soil moisture and crop evolution, and it can therefore be used as an indication of agricultural drought. The SPI-6 has a higher correlation with hydrological droughts, mainly represented by low anomalies in runoff. The SPI-12 and SPI-24 have a lower temporal variability and point to major and long-duration drought events whose impacts may extend to groundwater. The widely used SPI does, however, have its limitations mainly because it is based only on precipitation data. An extension of the SPI was proposed by Vicente-Serrano et al. (2010b), called the Standardized Precipitation Evaporation Index (SPEI), which is based on precipitation and potential evaporation. In a way, it combines the sensitivity of the PDSI to changes in evaporation demand with the capacity of the SPI to represent droughts on multi-temporal scales (Vicente-Serrano et al., 2010b).

Together with the development of the first drought indicators, hydrological models were also used for agricultural and hydrological drought assessment. Schulze (1984) applied the Agricultural Catchments Research Unit (ACRU) hydrological model in Natal, South Africa, to compare the severity of the 1979-1983 drought with other drought events in the previous 50 years. He identified hydrological modelling as a potentially powerful tool in drought assessment. Moreover, he indicated that it is necessary to distinguish between different types of droughts, as droughts in terms of water resources do not necessarily coincide with droughts from the crop production point of view.

In recent years, several new indicators have been developed to characterise the different types of drought. Although drought indicators are mostly used to characterise past droughts and monitor current droughts, forecasting of these indicators at different spatial and temporal scales is gaining considerable attention.

In this study we extend a continental-scale framework for drought forecasting in Africa, which is currently under development (Barbosa et al., 2013), and apply this to the Limpopo Basin in southern Africa, one of the most water-stressed basins in Africa. The Limpopo River basin is expected to face even more serious water scarcity issues in the future, limiting economic development in the basin (Zhu and Ringler, 2012). To apply this framework at the regional scale, a finer-resolution version of the global hydrological model PCRaster Global Water Balance (PCR-GLOBWB) was adapted to regional conditions in the basin. We model hydrological droughts and their space-time variability using a process-based distributed hydrological model in the (semi-) arid Limpopo Basin. The model was tested by comparing the simulated hydrological and agricultural drought indicators in the period 1979-2010 with reported historic drought events in the same period. We derive a number of different drought indicators from the model results (see Table 4-1), such as the ETDI (Evapotranspiration Deficit Index; Narasimhan and Srinivasan, 2005), the RSAI (Root Stress Anomaly Index), the SRI (Standardized Runoff Index; Shukla and Wood, 2008), and the GRI (Groundwater Resource Index; Mendicino et al., 2008). While the SRI is based on river discharge at a particular river section, the ETDI, RSAI and GRI are spatial indicators that can be estimated for any location in the basin. The ETDI and RSAI are directly related to water availability for vegetation with or without irrigation, and the GRI is related to groundwater storage. Moreover, we compute the widely known meteorological drought indicators SPI and SPEI at different aggregation periods to verify the correlation of the different aggregation periods for these indicators and the different types of droughts. Table 4-1 presents the derived indicators with a description of the purpose and the type of drought each indicator represents. The aim of this study is to assess the ability of different drought indicators to reconstruct the history of droughts in a highly water-stressed, semi-arid basin. Moreover, we investigate whether widely used meteorological indicators for drought identification can be complemented with indicators that incorporate hydrological processes.

Table 4-1 Drought indicators derived in this study.

Name	Variable	Type of drought	Purpose or reason	Reference
SPI (Standardised Precipitation Index)	Precipitation	Meteorological	Particularly important for rainfed agriculture; also influences farming practises	McKee et al. (1993)
SPEI (Standardised Precipitation Evaporation Index)	Precipitation/ Evaporation	Meteorological	As SPI, but with a more detailed focus on available water	Vicente-Serrano et al. (2010b)
ETDI (Evapotranspiration Deficit Index)	Evaporation	Agricultural	Impact on yield as a result of water availability for evaporation	Narasimhan and Srinivasan (2005)
RSAI (Root Stress Anomaly Index)	Root stress	Agricultural	Impacts on root growth and yield	This study
SRI (Standardized Runoff Index)	Discharge	Hydrological	River discharge is important for many aspects such as shipping, irrigation, energy	Shukla and Wood (2008)
GRI (Groundwater Resource Index)	Groundwater	Hydrological	Groundwater is used for irrigation and drinking water	Mendicino et al. (2008)

4.2 Data

4.2.1 Study area: Limpopo River basin

The Limpopo River basin has a drainage area of approximately 415,000 km² and is shared by four countries: South Africa (45%), Botswana (20%), Mozambique (20%) and Zimbabwe (15%) (Figure 4-1). The climate in the basin ranges from tropical dry savannah and hot dry steppe to cool temperatures in the mountainous regions. Although a large part of the basin is located in a semi-arid area, the upper part of the basin is located in the Kalahari Desert where it is particularly arid. Aridity, however, decreases further downstream. Rainfall in the basin is characterised as being seasonal and unreliable causing frequent droughts, but floods can also occur in the rainy season. The average annual rainfall in the basin is approximately 530 mm year⁻¹, ranging from 200 to 1,200 mm year⁻¹, and occurs mainly in the summer months (October to April) (LBPTC, 2010).

Arid and semi-arid regions are generally characterised by low and erratic rainfall, high interannual rainfall variability and a low rainfall-to-potential-evaporation ratio. This leads to the ratio of runoff to rainfall being low on the annual scale. Hydrological modelling possesses considerable challenges in such a region. A detailed discussion on problems related to rainfall-runoff modelling in arid and semi-arid regions can be found in Pilgrim et al. (1988).

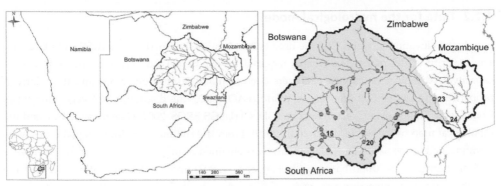

Figure 4-1 Limpopo River basin: the location of the basin (left panel) and the locations of hydrometric stations (right panel). Selected stations (nos. 1, 15, 18, 20, 23 and 24) are highlighted. The subbasins draining to each hydrometric station are named after the station number.

The runoff coefficient (RC = Runoff / Precipitation) of the Limpopo Basin is remarkably low. For the station at Chókwe (no. 24), which is the station with the largest drainage area among the discharge stations available in this study (Figure 4-1), the runoff coefficient is just 4.3% for the naturalised discharge and a mere 1.7% for the observed discharge (without naturalisation). Note that the naturalised discharge is estimated as observed discharge plus the estimated abstractions. These runoff coefficients are strikingly low: out of 539 mm year^{-1} of annual rainfall only 23 mm year^{-1} (basin average) turns into runoff annually including abstraction. This means that even a small error in estimates of precipitation and evaporation could result in a large error in the runoff. Moreover, the uncertainty in the rainfall input could easily be larger than the runoff coefficient (4.3%) of the basin. Runoff coefficients for other selected stations in the basin (highlighted in Figure 4-1, right panel) are presented in Table 4-2.

Table 4-2 Naturalised runoff coefficient (RCnat) and observed runoff coefficient (RCobs) for selected stations.

Station number	Subbasin area (km²)	Number of years without missing data	Mean annual observed runoff (m³ s^{-1})	RCnat (%)	RCobs (%)
24	342,000	27	96.9	4.3	1.7
23	259,436	26	82.1	3.8	2.0
1	201,001	17	39.5	3.0	1.2
18	98,240	29	12.2	3.6	0.7
20	12,286	24	14.8	6.3	5.3
15	7,483	32	4.6	6.3	3.1

The basin is also highly modified, as is evident from the observed and naturalised runoff. This adds an additional challenge to modelling this basin. For example, for the largest drainage outlet available (no. 24), the observed annual discharge is only some 39% of the naturalised discharge, which means that the abstractions in the basin amount to 61% of the total runoff. Irrigation water demand takes up the largest share. The total estimated present demand in the basin is about 4,700 x10⁶m³year^{-1}. The total natural runoff generated from rainfall is approximately 7,200 x10⁶m³year^{-1}, showing that a significant portion of the runoff generated in the basin is currently used.

4.2.2 Data for the hydrological model

The digital elevation model (DEM) we used is based on the Hydro1k Africa (USGS EROS, 2006). The majority of the parameters (maps) required for the model (soil layer depths, soil storage capacity, hydraulic conductivity, etc.) were derived mainly from three maps and their derived properties: the Digital Soil Map of the World (FAO, 2003), the distribution of vegetation types from Global Land Cover Characterization (GLCC) (USGS EROS, 2002; Hagemann, 2002), and the lithological map of the world (Dürr et al., 2005). From the soil map, 73 different soil types were distinguished in the basin. The irrigated area was obtained from the global map of irrigated areas in 5 arc-minutes resolution based on Siebert et al. (2007) and FAO (1997). We computed the monthly irrigation intensities per grid cell using the irrigated area map, the irrigation water requirement data per riparian country in the basin and the irrigation cropping pattern zones (FAO, 1997).

All meteorological forcing data used (precipitation, daily temperature, daily minimum and maximum temperature at 2 meter) are the same as in Section 3.2.1.1 and are based on the ERA-Interim (ERAI, Dee et al., 2011) reanalysis data set from the European Centre for Medium-Range Weather Forecasts (ECMWF). This data set covers the period from January 1979 to the present day with a horizontal resolution of approximately 0.7 degrees and 62 vertical levels. A comprehensive description of the ERAI product is available in Dee et al. (2011). The ERA-Interim precipitation data used with the present model were corrected with GPCP v2.1 (product of the Global Precipitation Climatology Project) to reduce the bias with measured products (Balsamo et al., 2010) as explained in Section 3.2.1.1. The GPCP v2.1 data are the monthly climatology provided globally at a 2.5° × 2.5° resolution, covering the period from 1979 to September 2009. Temperature data is used for the computation of the reference potential evaporation needed to force the hydrological model. In this study the Hargreaves formula was used. This method uses only temperature data (minimum, maximum and average), so it requires less parameterisation than Penman-Monteith, with the disadvantage that it is less sensitive to climatic input data, with a possibly reduction of dynamics and accuracy. However, it leads to a notably smaller sensitivity to error in climatic inputs (Hargreaves and Allen, 2003). Moreover, the potential evaporation derived from the Penman-Monteith equation and Hargreaves equation result in very similar values throughout Africa, and the choice of the method used for the computation of the reference potential evaporation appears to have minor effects on the results of the actual evaporation for southern Africa (Chapter 3, Trambauer et al., 2014a) . For this study, the ERAI data were obtained for the period of 1979-2010. These were converted to the same spatial resolution as the model using bilinear interpolation to downscale from the ERAI grid to the 0.5° model grid. ERAI is archived using an irregular grid (reduced Gaussian) over the domain and thus an interpolation was inevitable to be able to use it in the model.

Runoff data were obtained from the Global Runoff Data Centre (GRDC; http://grdc.bafg.de/), the Department of Water Affairs in the Republic of South Africa and ARA-Sul (Administração Regional de Águas do Sul, Mozambique). Runoff stations that had data available up until recent years, with relatively few missing data, are presented in Figure 4-1. Most of these stations are in the South African part of the basin as almost no data could be found from stations in the other riparian countries. The subbasins draining to each hydrometric station are named after the station number.

4.3 Methods

4.3.1 Process-based distributed hydrological model

4.3.1.1 General description

A process-based distributed hydrological (water balance) model based on PCR-GLOBWB (van Beek and Bierkens, 2009) is used. First the global-scale model was adapted to the continent of Africa (Chapter 3, Trambauer et al., 2014a). A higher-resolution version (0.05° x 0.05°) of the continental model (0.5° x 0.5°) was applied for the Limpopo River basin. The PCR-GLOBWB was one of the 16 different land surface and hydrological models reviewed (Chapter 2, Trambauer et al., 2013), and it was identified as one of the hydrological models that can potentially be used for hydrological drought studies in large river basins in Africa. PCR-GLOBWB is in many ways similar to other global hydrological models, but it has many improved features, such as improved schemes for sub-grid parameterisation of surface runoff, interflow and baseflow, a kinematic-wave-based routing for the surface water flow, dynamic inundation of floodplains, and a reservoir scheme (van Beek and Bierkens, 2009; van Beek, 2008).

On a cell-by-cell basis and at a daily time step, the model computes the water storage in two vertically stacked soil layers (max. depth 0.3 and 1.2 m) and in an underlying groundwater layer, as well as computing the water exchange between the layers and between the top layer and the atmosphere. It also calculates canopy interception and snow storage. Within a grid cell, the sub-grid variability is taken into account considering tall and short vegetation, open water and different soil types. Crop factors are specified on a monthly basis for short- and tall-vegetation fractions, as well as for the open-water fraction within each cell, as described in Section 3.2.1.1. The total specific runoff of a cell consists of the surface runoff (saturation excess), snowmelt runoff (after infiltration), interflow (from the second soil layer) and baseflow (from the lowest reservoir as groundwater). River discharge is calculated by accumulating and routing specific runoff along the drainage network and including dynamic storage effects and evaporative losses from lakes and wetlands (van Beek and Bierkens, 2009; van Beek, 2008). The default PCR-GLOBWB model does not explicitly consider irrigated areas but the version of the model used here includes an irrigation module to account for the highly modified hydrological processes in the irrigated areas of the basin.

4.3.2 Drought indicators

The meteorological drought indicators used in this study are computed only from meteorological variables: precipitation and potential evaporation. Agricultural and hydrological indicators, on the other hand, are computed from the results of the hydrological model, and therefore account for effects of soil, land use, groundwater characteristics, etc. in the basin. The indicators used in this study are described below.

4.3.2.1 Meteorological drought indicators

Standardised Precipitation Index (SPI)

The SPI was developed by McKee et al. (1993) and it interprets rainfall as a standardised departure with respect to a rainfall probability distribution. It requires fitting the precipitation time series to a gamma distribution function, which is then transformed to a normal distribution allowing the comparison between different locations. The SPI [-] is then computed as the discrete precipitation anomaly of the transformed data divided by the standard deviation of the transformed data (Keyantash and Dracup, 2002; McKee et al., 1993). SPI values mainly range from 2.0 (extremely wet) to -2.0 (extremely dry).

Standardised Precipitation Evaporation Index (SPEI)

Instead of using only precipitation as in the SPI, the SPEI uses the difference between precipitation (P) and potential evaporation (PET), i.e. $D = P\text{-}PET$, and the PET is computed following the Thornthwaite method (Vicente-Serrano et al., 2010a; Vicente-Serrano et al., 2010b). The calculated D values are aggregated at different timescales, following the same procedure as for the SPI. A log-logistic probability function is then fitted to the data series of D and the function is then standardised following the classical approximation of Abramowitz and Stegun (1965). The SPEI also ranges between -2.0 and 2.0; the average value of the SPEI is 0, and the standard deviation is 1 (Vicente-Serrano et al., 2010a; Vicente-Serrano et al., 2010b).

4.3.2.2 Agricultural drought indicators

Agricultural droughts are defined as the lack of soil moisture to fulfil crop demands, and therefore the agriculture sector is normally the first to be affected by a drought. In this study we characterise agricultural droughts by means of two spatially distributed indicators defined as described in the following.

Evapotranspiration Deficit Index (ETDI)

The ETDI (Narasimhan and Srinivasan, 2005) is computed from the anomaly of water stress to its long-term average. The monthly water stress ratio (WS [0-1]) is computed as

$$WS = \frac{PET-AET}{PET} \qquad (4\text{-}1)$$

where PET and AET are the monthly reference potential evaporation and monthly actual evaporation, respectively. The monthly water stress anomaly (WSA) is calculated as

$$WSA_{y,m} = \frac{MWS_m-WS_{y,m}}{MWS_m-minWS_m} \times 100, \; if \; WS_{y,m} \leq MWS_m \qquad (4\text{-}2)$$

$$WSA_{y,m} = \frac{MWS_m-WS_{y,m}}{maxWS_m-MWS_m} \times 100, \; if \; WS_{y,m} > MWS_m \qquad (4\text{-}3)$$

where MWS_m is the long-term median of water stress of month m, $maxMWS_m$ is the long-term maximum water stress of month m, $minWS_m$ is the long-term minimum water stress of month m, and $WS_{y,m}$ is the monthly water stress ratio (y= 1979-2010 and m= 1-12). Narasimhan and Srinivasan (2005) scaled the ETDI to between -4.0 and 4.0 to be comparable with the PDSI. Here, we used the same scaling procedure but amended this to scale the ETDI to between -2.0 and 2.0 to make it comparable to the SPI, SPEI and SRI

$$ETDI_{y,m} = 0.5ETDI_{y,m-1} + \frac{WSA_{y,m}}{100} \qquad (4\text{-}4)$$

Root Stress Anomaly Index (RSAI)

The 'root stress' (RS) is a spatial indicator of the available soil moisture, or the lack of it, in the root zone. The root stress varies from 0 to 1, where 0 indicates that the soil water availability in the root zone is at field capacity and 1 indicates that the soil water availability in the root zone is at wilting point and the plant is under maximum water stress. The RSAI is computed similarly to the ETDI described above. The monthly root stress anomaly (RSA) is calculated as

$$RSA_{y,m} = \frac{MRS_m - RS_{y,m}}{MRS_m - minRS_m} \times 100, \; if \; RS_{y,m} \leq MRS_m \qquad (4\text{-}5)$$

$$RSA_{y,m} = \frac{MRS_m - RS_{y,m}}{maxRS_m - MRS_m} \times 100, \; if \; RS_{y,m} > MRS_m \qquad (4\text{-}6)$$

where MRS_m is the long-term median root stress of month m, $maxMRS_m$ is the long-term maximum root stress of month m, $minRS_m$ is the long-term minimum root stress of month m, and $RS_{y,m}$ is the monthly root stress (y= 1979-2010 and m= 1-12). The root stress anomaly index, scaled to between -2.0 and 2.0 (using the same procedure as Narasimhan and Srinivasan, 2005) is

$$RSAI_{y,m} = 0.5RSAI_{y,m-1} + \frac{RSA_{y,m}}{100} \qquad (4\text{-}7)$$

4.3.2.3 Hydrological drought indicators

For the characterisation of hydrological droughts we used the commonly applied Standardised Runoff Index (SRI; Shukla and Wood, 2008) for streamflow and the Groundwater Resource Index (GRI; Mendicino et al., 2008) for groundwater storage.

Standardised Runoff Index (SRI)

The SRI follows the same concept as the SPI and is defined as a "unit standard normal deviate associated with the percentile of hydrologic runoff accumulated over a specific duration" (Shukla and Wood, 2008). To compute the SRI the simulated runoff time series is fitted to a probability density function (a gamma distribution is used here), and the function is used to estimate the cumulative probability of the runoff of interest for a specific month and temporal scale. The cumulative probability is then transformed to the standardised normal distribution with a mean of 0 and a variance of 1 (Shukla and Wood, 2008).

Groundwater Resource Index (GRI)

The $GRI_{y,m}$ is suggested as a standardisation of the monthly values of groundwater storage (detention) without any transformation (Mendicino et al., 2008):

$$GRI_{y,m} = \frac{S_{y,m} - \mu_{S,m}}{\sigma_{S,m}} \qquad (4\text{-}8)$$

where $S_{y,m}$ is the value of the groundwater storage for the year y and the month m, and $\mu_{S,m}$ and $\sigma_{S,m}$ are respectively the mean and the standard deviation of the groundwater storage S simulated for the month m in a defined number of years (32 years in this case). The same classification that is used for the SPI (between -2.0 and 2.0) is applied to the GRI (Wanders et al., 2010).

4.3.3 Identification of past droughts and primary characterisation of drought severity

To identify past droughts, the drought indicators described were calculated for the period 1979-2010 for the Limpopo River basin, resulting in times series of monthly indicator maps. The maps allow for the visualisation of the spatial variability of the indicators in the basin. The SPI, SPEI and SRI were computed for different aggregation periods (1, 3, 6, 12 and 24). All the indicators were then aggregated over several subbasins resulting in times series for each indicator. The historical subbasin-averaged indicators were then compared. Maps of the indicators are also compared for specific years to show the spatial variability of the indicators and the extent of the droughts.

All indicators considered were scaled to range between -2.0 and 2.0. Based on the SPI values, droughts may be classified into mild ($0 \geq$ SPI > -1.0), moderate ($-1.0 \geq$ SPI > -1.5), severe ($-1.5 \geq$ SPI > -2.0) and extreme (SPI ≤ -2.0) (Lloyd-Hughes and Saunders, 2002; see Table 4-3). For the SPI and SPEI, the spatially averaged indicators are no longer related to a probability of occurrence. However, we still use the same thresholds for the characterisation of the subbasin aggregated droughts, as we understand that the resulting indicators would not be very different from the computation of these indicators with aggregated precipitation and potential evaporation. For agricultural (ETDI and RSAI) and groundwater indicators (GRI) this is not the case as these are not defined based on a probability of occurrence.

Table 4-3 State definition according to the indicator value.

Indicator value (Iv)	State category
Iv > 2.0	Extremely wet
1.5 < Iv ≤ 2.0	Severely wet
1.0 < Iv ≤ 1.5	Moderately wet
0 < Iv ≤ 1.0	Mildly wet
-1.0 < Iv ≤ 0	Mild drought
-1.5 < Iv ≤ -1.0	Moderate drought
-2.0 < Iv ≤ -1.5	Severe drought
Iv ≤ -2.0	Extreme drought

Droughts are generally characterised by a start date and an end date (both defining duration), drought intensity (indicator value), and severity or drought magnitude. The drought severity (DS) definition by McKee et al. (1993) is used here:

$$DS = -\left(\sum_{j=1}^{x} Iv_{ij}\right) \tag{4-9}$$

Where Iv is the indicator value; j starts with the first month of a drought and continues to increase until the end of the drought (x) for any of the i timescales. The DS (months) would be numerically equivalent to the drought duration if the drought had an intensity (value) of -1.0 for each month (McKee et al., 1993).

4.4 Results and discussion

4.4.1 Hydrological model performance

Given the complexity of the basin for hydrological modelling, particularly due to the arid or semi-arid nature, the model results are quite satisfactory, especially for the larger subbasins. Runoff estimates from the hydrological model were verified with observed runoff on a monthly basis. For a number of the runoff stations tested, the coefficient of determination (R^2) values varied from about 0.45 to as good as 0.92. In a review of model application and evaluation, Moriasi et al. (2007) recommended three quantitative statistics for model evaluation: Nash-Sutcliffe efficiency (NSE), percent bias (PBIAS) and the ratio of the root mean square error to the standard deviation of the measured data (RSR). They also specified ranges for these statistics for a "satisfactory" model performance (NSE > 0.5, RSR ≤ 0.70, and PBIAS ± 25% for streamflow). However, PBIAS is highly influenced by uncertainty in the observed data (Moriasi et al., 2007). Given the potential problems in observed flow data in South Africa reported by the Water Research Commission (2009) such as poor accuracy of the rating table, particularly at low flows and the inability to measure high flows, we do not evaluate our results based on PBIAS. The evaluation measures NSE and RSR together with the coefficient of determination for selected stations are presented in Table 4-4. We do not calibrate parameters based on these evaluation measures, but we use them as a simple test of concordance. Based on the ranges proposed by Moriasi et al. (2007), the model performance is found to be satisfactory for four out of six runoff stations.

Table 4-4 Model evaluation measures for runoff for selected stations.

Station number	R^2	NSE	RSR
24	0.92	0.90	0.32
23	0.62	0.38	0.79
1	0.69	0.57	0.65
18	0.68	0.62	0.62
20	0.70	0.65	0.59
15	0.53	0.48	0.72

4.4.2 Identification of historic hydrological droughts in the basin

Drought indicators were computed for the period 1979-2010. Agricultural and hydrological drought indicators were computed from the fluxes resulting from the hydrological model. Because the focus in the current model is to simulate hydrological droughts, it is important that the model captures the most important drought events in the simulation period 1979-2010. DEWFORA (2012a) reported that in the period 1980-2000, the southern African region was struck by four major droughts, notably in the seasons 1982/83, 1986/87, 1991/92 and 1994/95. The drought of 1991/92 was the most severe in the region in recent history. After the year 2000, important droughts include the years 2002/03/04 and 2005/06. Droughts in the Limpopo River basin also show significant spatial variability. A study covering only the Botswana part of the basin documents a severe drought that occurred in 1984 (Dube and Sekhwela, 2007). However, in that year no documentation of drought in the other parts of the basin was found.

4.4.2.1 Agricultural droughts

Figure 4-2 presents the RSAI and ETDI for the most severe drought in recent history (1991/92), for the very dry year 1982/83, for a wet year (1999/2000) and for a year with both dry and wet conditions at different locations in the basin (1984/85). The geographic variability of the RSAI seems to be slightly higher than that of the ETDI. These indicators provide information for the assessment of agricultural droughts. The figure shows that both indicators, computed from different outputs of the hydrological model (actual evaporation and soil moisture), produce similar results and are able to reproduce the dry or wet conditions in the basin. This is also supported by Figure 4-3, which shows the fraction of the Limpopo Basin under moderate to extreme agricultural drought, i.e. Iv ≤ -1.0. Both indicators illustrate that a large part of the basin was under at least moderate agricultural drought conditions for the years with recorded drought events.

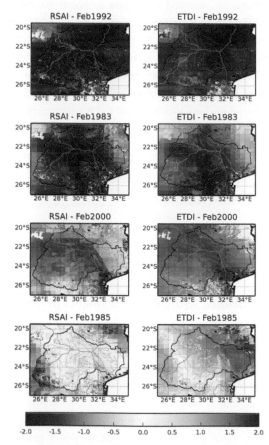

Figure 4-2 Root Stress Anomaly Index (RSAI) and Evapotranspiration Deficit Index (ETDI) in the Limpopo Basin for selected years.

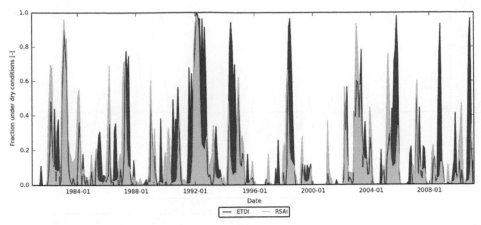

Figure 4-3 Fraction of the Limpopo Basin under moderate to extreme droughts represented by the indicator value (Iv ≤ -1.0).

4.4.2.2 Hydrological droughts

Figure 4-4 shows the SRI values (1-month, 3-months, 6-months and 12-months) from 1979 to 2010 computed from the simulated runoff at station 24. The dotted grey line at the threshold value of -1.0 is used to identify moderate droughts, with the moderate drought considered to start when the indicator falls below the threshold, and stop when the indicator goes above the threshold. The simulated SRI clearly identifies the severe hydrological droughts of 1982/83 and 1991/92 and the very wet (flood) year of 1999/2000. The SRI from observed data was not included in the figure given that there are periods with missing data and the computation of the SRI requires a monthly runoff data set for a continuous period without missing data.

The Groundwater Resource Index (GRI) presented in Figure 4-5 for the same selected years shows the years 1991/92 and 1982/83 to be drier than normal, but the intensity of the drought appears to be quite low (not severe). The year 1984/85, selected as it presents both dry and wet conditions at different locations in the basin, does not show this spatial variability for the GRI. This was to be expected due to the persistence of the groundwater storage and low intensity of indicators of drought or wetness in this year in different locations of the basin. The intensity of the extremely wet year 1999/2000 is well represented, suggesting that the GRI is skewed. This is likely due to the fact that the GRI is not transformed into the normal space. Moreover, the distribution of values is constrained by the capacity of the groundwater reservoir in the hydrological model. Mendicino et al. (2008) applied this indicator in a Mediterranean climate but the skewness test of normality showed that their series from January to September were normally distributed, while the series of October to December were not normally distributed. However, they indicate that the values of groundwater storage in the last winter months and in spring were more important. For this indicator to be applied independently of the climate and basin conditions it should probably be transformed into the normal space.

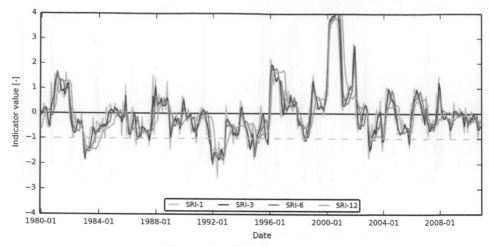

Figure 4-4 Simulated SRI for station 24.

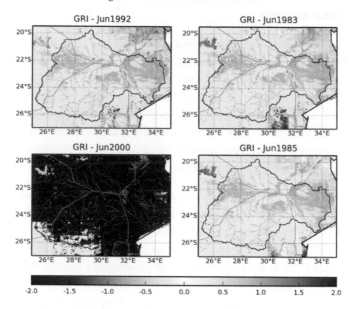

Figure 4-5 Groundwater Resource Index (GRI) for selected years.

4.4.2.3 Comparison of drought indicators

The computed indicators were averaged for the whole basin as well as for the selected subbasins. Time series of the resulting indicators were compared for the whole 1980-2010 period. Figure 4-6 presents the time series of aggregated drought indicators for subbasin 24. Note that the subbasins are named after the hydrometric station number. Figure 4-6 compares the agricultural, hydrological and groundwater drought indicators. The agricultural indicators ETDI and RSAI are compared with the meteorological drought indicators SPI and SPEI with the short aggregation period (3 months) that is commonly used as indicators of agricultural droughts. Figure 4-6 (upper plot) shows that the indicators are mostly in phase, correctly representing the occurrence of dry and wet years, and the intensities of the events are in general quite similar. The hydrological drought indicator SRI-6 is compared with the meteorological drought indicators SPI-6 and SPEI-6 (upper middle plot). All three indicators follow roughly the same pattern, but the fluctuation of the SRI seems to be slightly lower than that of the meteorological indices (SPI and SPEI). This is probably due to the higher persistence of streamflow when compared to precipitation. Moreover, it is clearly visible from Figure 4-6 that the temporal variability or fluctuation of the indicators reduces when moving from drought indicators associated with agricultural drought to those associated with hydrological drought. This means that several mild agricultural droughts do not progress further to hydrological droughts.

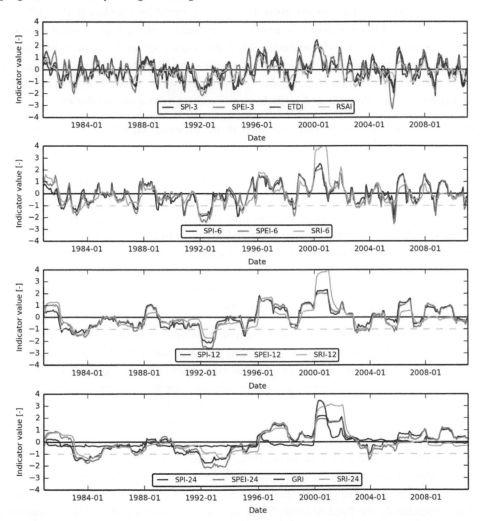

Figure 4-6 Time series of aggregated drought indicators for subbasin 24. Upper graph: Indicators used to characterise agricultural droughts (SPI-3, SPEI-3, ETDI and RSAI); upper middle graph: indicators used to characterise hydrological drought (SPI-6, SPEI-6 and SRI-6); lower middle graph: indicators used to characterise groundwater droughts (SPI-12, SPEI-12 and SRI-12); and lower graph: indicators used to characterise extended groundwater droughts (SPI-24, SPEI-24, GRI and SRI-24).

Moreover, to identify groundwater droughts, or major drought events, the time series of the GRI is compared to the time series of meteorological drought indicators with long aggregation periods (SPI-12, SPEI-12, SRI-12, SPI-24, SPEI-24, SRI-24) (see Figure 4-6, lower middle and lower plots). The plots show that as the variability of the indicator reduces further the number of multi-year prolonged droughts increases. However, for groundwater droughts only two events (1982/83 and 1991/92) are identified as moderate to severe droughts ($Iv \leq -1.0$). The plots again show that, in general, the temporal variability of the runoff-derived indicator (SRI) is lower than that of the meteorological indicators (SPI and SPEI). The GRI shows much less temporal variability than the other indices and does not identify any extreme events, with the exception of the flood of 1999/2000. Similar results using the GRI were found by Wanders et al (2010), who indicate that the GRI has a very low number of droughts with a high average duration. Moreover, a study of Peters and Van Lanen (2003) investigated groundwater droughts for two climatically contrasting regimes. For the semi-arid regime they found multi-annual droughts to occur frequently. They indicate that the effect of the groundwater system is to pool erratically occurring dry months into prolonged groundwater droughts for the semi-arid climate.

Table 4-5 presents a correlation matrix between all the indicators considered in this study for subbasin 24. Similar correlation results were found for the other subbasins. The table shows that the agricultural drought indicators ETDI and RSAI have the highest correlation with the SPEI-3, SPEI-6, SPI-3, SPI-6 and with the SRI with low aggregation periods (1 to 3 months). For every station the correlation between the agricultural indicators and the SPEI is slightly higher than with the SPI. While the hydrological drought indicators SRI-6 and SRI-12 present the highest correlation with the meteorological drought indicators SPI-12 and SPEI-12, the extended hydrological drought indicator SRI-24 is better correlated with the meteorological drought indicators SPI-24 and SPEI-24. The GRI shows the highest correlation with the SRI-6 and SRI-12. This makes sense given the direct connection between groundwater and runoff, where groundwater (baseflow) contributes to the total runoff.

Figures 4-7, 4-8, and 4-9 present the monthly spatial mean time series of drought indicators for subbasins 1, 18 and 20, respectively. The averaged indicators for subbasins 24 and 1, the two largest subbasins considered, are almost identical (see Figure 4-6 and Figure 4-7). Figure 4-8 shows that even though the general pattern of the time series for the subbasin 18 is similar to that found for subbasin 24 and 1, some differences are noticeable. For example, Figure 4-8 shows a clear drought period for subbasin 18 in the years 1984/85/86 which is not apparent for the subbasins 24 and 1. These localised drought events hat affected the upper part of the basin were not apparent for the lower part of the basin. This was also observed in Figure 4-2. Moreover, the extreme floods that occurred in the lower part of the basin in 1999/2000 are much less severe in the upstream parts of the basin. For example, Figure 4-9 shows that for subbasin 20 (the smallest subbasin considered), the flood of 1996/97 was more severe than that of 1999/2000. Similarly, while the drought of 2003/04 is quite mild when averaged over the largest selected subbasin (no. 24), it is quite severe for subbasin 20 (similar to the droughts of 1983/84 and 1991/92).

Table 4-5 Correlation matrix of drought indicators for subbasin 24. Bold italic numbers indicate correlations higher than 0.7.

	SPI-3	SPEI-3	ETDI	RSAI	SPI-6	SPEI-6	SRI-1	SRI-2	SRI-3	SRI-6	SPI-12	SPEI-12	SRI-12	SPI-24	SPEI-24	SRI-24	GRI
SPI-3	1.00																
SPEI-3	0.91	1.00															
ETDI	0.79	0.82	1.00														
RSAI	0.70	0.73	0.84	1.00													
SPI-6	0.77	0.75	0.81	0.79	1.00												
SPEI-6	0.72	0.80	0.83	0.80	0.94	1.00											
SRI-1	0.69	0.71	0.84	0.84	0.74	0.75	1.00										
SRI-2	0.67	0.71	0.83	0.85	0.75	0.78	0.97	1.00									
SRI-3	0.63	0.68	0.81	0.85	0.75	0.78	0.94	0.99	1.00								
SRI-6	0.51	0.58	0.75	0.82	0.71	0.76	0.87	0.93	0.96	1.00							
SPI-12	0.53	0.56	0.70	0.74	0.73	0.75	0.72	0.75	0.78	0.82	1.00						
SPEI-12	0.48	0.58	0.68	0.71	0.68	0.77	0.69	0.73	0.75	0.81	0.96	1.00					
SRI-12	0.37	0.45	0.61	0.65	0.53	0.61	0.74	0.80	0.83	0.91	0.81	0.83	1.00				
SPI-24	0.45	0.48	0.56	0.59	0.60	0.63	0.65	0.68	0.69	0.72	0.79	0.80	0.76	1.00			
SPEI-24	0.40	0.48	0.52	0.54	0.54	0.62	0.60	0.62	0.64	0.67	0.71	0.79	0.73	0.96	1.00		
SRI-24	0.30	0.35	0.44	0.46	0.42	0.47	0.60	0.64	0.67	0.71	0.57	0.62	0.80	0.83	0.84	1.00	
GRI	0.37	0.34	0.48	0.57	0.48	0.42	0.72	0.75	0.76	0.76	0.57	0.48	0.73	0.60	0.49	0.66	1.00

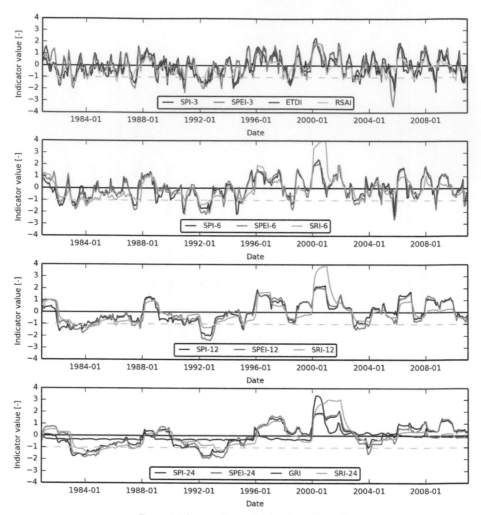

Figure 4-7 Same as Figure 4-6, but for subbasin 1.

For the four subbasins a short but intense agricultural drought is noticeable at the beginning of the 2005/06 season, but this did not progress to an extended hydrological drought. This is consistent with the literature, which indicates that this season was delayed and after a dry start to the season, good rainfall occurred from the second half of December (Department of Agriculture of South Africa, 2006). In subbasin 18 (Figure 4-8), even though meteorological indicators (SPI-6, SPEI-6, SPI-12 and SPEI-12) suggest that the 1986/87 season was near normal to wet, the hydrological indicators (SRI-6, SRI-12) point to a dry runoff year. Measured runoff at this station indicates that the year 1986/87 was indeed a dry year. This seems similar to what was found by Peters and van Lanen (2003); for longer aggregations periods an accumulation of successive short anomalies can lead to an overall hydrological drought. Similarly, meteorological indicators suggest that the floods of 1996/97 and 1999/2000 in the lower part of the basin were of a similar magnitude. However, records indicate that the flood of 1999/2000 was much more extreme than the one of 1996/97 (WMO, 2012a). This can be seen clearly in the hydrological drought indicators

SRI-6, SRI-12 and SRI-24. The GRI shows almost no departure from normal, with the exception of the flood of 1999/2000. These results show the importance of computing indicators that can be related to hydrological drought and how these add value to the identification of droughts or floods and their severity. The indicators also help identify the spatial and temporal evolution of drought and flood events that would otherwise not have been apparent when considering only meteorological indicators.

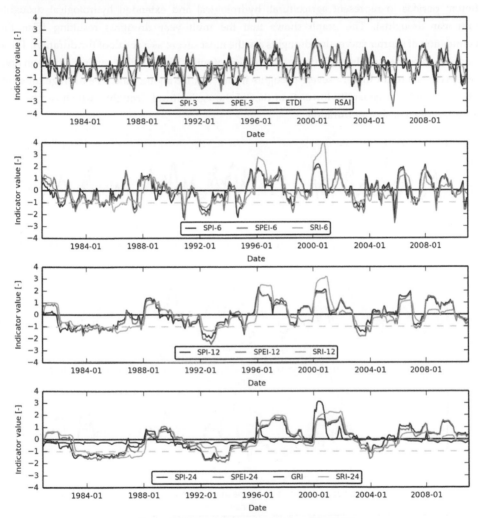

Figure 4-8 Same as Figure 4-6, but for subbasin 18.

We also computed drought severities (DS (months)) resulting from the different indicators as explained in section 4.3.3 (Eq. 4- 9). The droughts of 1982/83, 1986/87, 1991/92, 1994/95, 2002/03/04 and 2005/06 are identified as being among the most severe droughts, but the end month of these drought events varies for the different indicators. The indicators with higher aggregation periods (e.g 12 and 24 months), which have a lower temporal variability, generally point to longer droughts (multi-year droughts) with higher persistence than indicators with lower aggregation periods (agricultural droughts). For example, while the agricultural indicators suggest that the extreme drought of 1991/92 was over by the end of 1992 or beginning of 1993, the indicators that represent hydrological droughts signal that this drought only ended at the end of 1993. Moreover, for the SRI-12, GRI, SPI-24 and SPEI-24, this multi-year drought lasts until 1994/95. As an example for subbasin 24, Figure 4-10 presents the duration and severity of the six most severe recorded droughts as identified by the meteorological drought indicator SPEI aggregated for different periods to represent agricultural, hydrological and extended hydrological droughts (multi-year droughts). The graph shows that the multi-year droughts resulting from the accumulation of shorter successive droughts are the most severe as a result of the duration. These droughts can be the most hazardous, as a succession of mild droughts that can initially seem non-problematic can result in very severe droughts if they last for a long time. The average intensity of these droughts is generally lower than that of the agricultural droughts, which can be very intense but often of shorter duration.

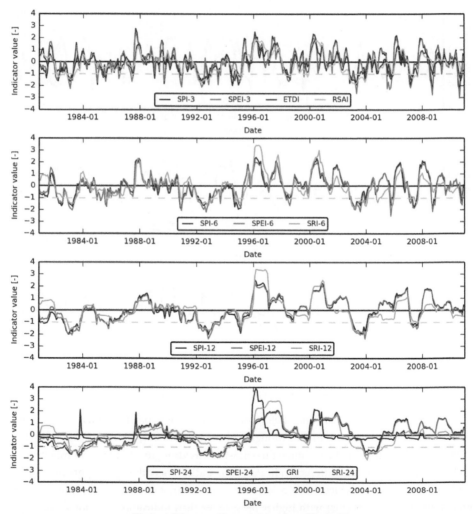

Figure 4-9 Same as Figure 4-6, but for subbasin 20.

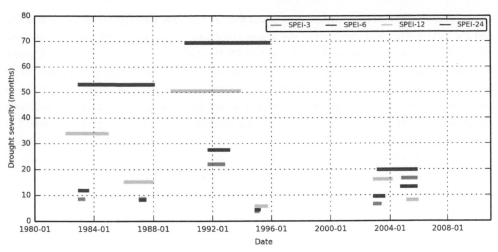

Figure 4-10 Drought severity and duration in subbasin 24 for the six most severe droughts in the period 1980-2010 for the indicator SPEI with different aggregation periods.

4.5 Conclusions

Very low runoff coefficients and high rainfall variability pose major challenges in modelling hydrological droughts in (semi-)arid basins. Small errors in the meteorological forcing and estimation of evaporation may result in significant errors in the runoff estimation. This also implies that model calibration, if any, should be applied cautiously to maintain the physical meaning of model parameters. We opted to apply a process-based model and parameterise it on the basis of the best available input data without additional calibration. In the process we ensured that we were using reliable data sets and interpolated or aggregated them with care to prepare spatially distributed parameter maps. The model is able to simulate hydrological drought-related indicators reasonably well. We have derived a number of different drought indicators from the model results, such as the ETDI, RSAI, SRI and GRI. While the SRI is based on river runoff at a particular river section, the ETDI, RSAI and GRI are spatial indicators that can be estimated at any location in the basin. The ETDI and RSAI are directly related to water availability for vegetation with or without irrigation, and the GRI is related to groundwater storage. Moreover, we computed the widely known drought indicators SPI and SPEI at different aggregation periods to verify the correlation of the different aggregation periods for these indicators and the different types of droughts.

All the indicators considered (with the exception of the GRI) are able to represent the most severe droughts in the basin and to identify the spatial variability of the droughts. Our results show that even though meteorological indicators with different aggregation periods serve to characterise droughts reasonably well, there is added value in computing indicators based on the hydrological model for the identification of droughts or floods and their severity. The indicators also help identify the spatial and temporal evolution of drought and flood events that would otherwise not have been apparent when considering only meteorological indicators.

The RSAI follows the ETDI to a great extent, and the ETDI is quite well represented by the SPEI-3 and the SPEI-6. This indicates that in the absence of actual evaporation and soil moisture data which are required to compute the ETDI and RSAI, the meteorological indicator SPEI-3, which considers both precipitation and potential evaporation and is reasonably easy to compute may be used as an indicator of agricultural droughts. For discharge we observe some added value in computing the SRI. Even though the SPI can give a reasonable indication of drought conditions, computing the SRI can be more effective for the identification of hydrological droughts. The groundwater indicator GRI mostly remains near normal conditions. A combination of different indicators, such as the SPEI-3, SRI-6 and SPI-12 (computed together), can be an effective way to characterise agricultural to long-term hydrological droughts in the Limpopo River basin.

5

DOWNSCALING THE OUTPUT OF A LOW RESOLUTION HYDROLOGICAL MODEL TO HIGHER RESOLUTIONS

This chapter discusses spatial scale in hydrological modelling based on the two resolution versions of the hydrological model presented in Chapters 3 and 4. These are a "low" resolution (0.5° × 0.5°) hydrological model, which covers the entire Africa, and a "high" resolution (0.05° × 0.05°) model for the Limpopo River basin. Using the results of the two models, we compute the effect of spatial resolutions on three distributed fluxes or storages: actual evaporation, soil moisture, and total runoff. Secondly, we examine the potential for downscaling of the low resolution hydrological model outputs to a higher resolution with a simple bias correction approach and a more comprehensive approach based on physical characteristic of the basin. Actual evaporation presents the lowest coefficient of variation (CV), followed by soil moisture, and total runoff presents the highest CV. Dry seasons and dry years are identified with higher CV than wet seasons and years. Moreover, results show that for the wet season, there is good potential of downscaling the low resolution hydrological model results to high resolution based on the terrain, soil and lithological characteristics.

This chapter is based on:

Trambauer, P., Maskey, S., Werner, M., Uhlenbrook, S. and van Beek, L.P.H: Downscaling the output of a low resolution hydrological model to higher resolutions, submitted to Hydrological Processes.

5.1 Introduction

Scaling is considered as one of the most challenging problem in hydrological modelling (Wigmosta and Prasad, 2006; Blöschl, 2001). Upscaling and downscaling methods are necessary to transfer information from small-scale data to large-scale predictions and vice versa. Viney and Sivapalan (2004) emphasized the need of large-scale conceptualizations of hydrological processes for the development of catchment scale rainfall-runoff models and for linking surface hydrology to global climate models. In recent years, an increasing number of global hydrological models have been developed by different research groups for a variety of purposes, e.g. PCR-GLOBWB (van Beek and Bierkens, 2009), LISFLOOD (van der Knijff and de Roo, 2008; De Roo et al., 2000) or VIC (Liang et al., 1994). This development has in part been due to developments in global climate models and remote sensing techniques that allow collecting a vast amount of distributed data covering the entire earth. Moreover, the advances in computing power also played a major role. In spite of the advances in computing power, computational time still somehow limits the resolution of these hydrological models. In addition, the spatial resolution of the hydrological model is often chosen to be commensurate to the spatial resolution of the global input data, most notably climate input. Thus, meteorological data from satellites, reanalysis data, or global data sets (such as GPCP, GPCC, ERA-40, ERA-Interim and CRU), with spatial resolution of 0.5° or lower, may be the best forcing available, and the only ones that are available over longer periods in data scarce areas. For this reason, global hydrological models in general have spatial resolution or grid size in the order of 0.5 degrees. However, results from these models are, in many cases, required at higher resolutions for a number of hydrologic applications, including agriculture and land management (Busch et al., 2012). The need of higher resolution global hydrological models is therefore recognized (Bierkens et al., 2015; Wood et al., 2011). Alternatively, it would be useful to downscale any of the resulting hydrological variables (e.g. evaporation, soil moisture, runoff, etc.) from these global hydrological models to finer spatial scales for any desired region in order to recapture the spatial variability of the processes that was smoothed out in the global models. For some purposes such as prediction of flooding extent, where an initial fast assessment of the situation is desired, downscaling of hydrological variables results would be much more efficient than running high resolution hydrological models due to the computation time gained. Recovering the spatial variability of smoothed out results in low resolution models is, however, extremely difficult, and initial approaches will remain empirical and site-specific.

There are two classical approaches or methods for upscaling and downscaling. The first involves dynamic models of parts of the hydrological cycle where the scaling issue is how the model equations and model parameters will change with scale. The second approach applies statistical descriptions where the focus is on how to best represent random variability in both space and time at diverse scales (Blöschl, 2006). Even though there is a vast amount of literature on upscaling and downscaling methods in the various subdisciplines of hydrology (Blöschl, 2006; Blöschl et al., 2009), downscaling is most often heard of in atmospheric sciences. Downscaling of hydrological variables is less frequently used, and is mostly limited to soil moisture (Blöschl, 2006; Western et al., 1999; Blöschl et al., 2009; Moore et al., 1988; Ladson and Moore, 1992; Busch et al., 2012). In contrast, upscaling of land surface and soil parameters, and climatological variables are widely cited (Stewart et al., 1996; Shuttleworth et al., 1997; Farmer, 2002).

Downscaling is a broadly used term in the field of atmospheric sciences, which Hewitson and Crane (1996) define as "a term adopted ... to describe a set of techniques that relate local- and regional-scale climate variables to the larger scale atmospheric forcing". In atmospheric sciences, the dynamical downscaling approach involves a higher resolution climate model embedded within a global circulation model (GCM). In the statistical approach, statistical methods are used to establish empirical relationships between GCM-resolution climate variables and local climate (Fowler et al., 2007). Similarly, establishing catchment- scale hydrological variables from the larger regional-scale hydrological ones is of interest for a number of applications. This is also often described as disaggregation of variables (Blöschl and Sivapalan, 1995; Bierkens et al., 2000).

Hydrological models are commonly categorised based on how they represent (i) the physics of the processes involved (conceptual, empirical or physically based), and (ii) the spatial discretisation or resolution (distributed or lumped). Whereas lumped model do not represent spatial variability explicitly, distributed models account for spatial variability at the defined scale by discretisising the landscape into grid cells. The problem of scaling can be viewed differently in different types of models. Wigmosta and Prasad (2006) argue that the empirical models should not be scaled beyond the original range of development, while conceptual models may represent more or fewer phenomena as the scale changes, and physically-based distributed models are confronted with the problem of variability in input variables as the resolution changes. In some physically based models, the equations concerning the various hydrological processes are solved for each grid cell, and the heterogeneity of the hydrological quantities inside a grid cell are ignored (Pradhan et al., 2006). However, many global and macro-scale hydrological models (e.g. PCR-GLOBWB) take this into account by using sub-grid variability.

In physically-based distributed models it is generally easier to upscale by aggregating outputs obtained at finer resolution than to downscale from a coarse resolution. When upscaling it is clear that the variability of the signals present at high resolution will be reduced, while downscaling presents the difficulty of disaggregating the variability that was not present at the original scale. In the models that consider sub-grid variability, the jump between two scales may be not so extreme. In some cases, scaling may require a different mathematical representation of a particular process, and therefore the original model cannot be simply scaled but must be reformulated to include previously neglected processes (Wigmosta and Prasad, 2006; Blöschl and Sivapalan, 1995). The problem of how to integrate the governing equations over the grid cell is a central issue regardless of the types of models (Martina and Todini, 2008). Martina and Todini (2008) show that the TOPKAPI hydrological model produced acceptable results upon a grid scale of the order of a kilometre. Beyond that scale, they find greater divergences of the subsurface flow from the correct solution. They indicate that a correct integration of the differential equations across various scales can generate relative scale independent models, which preserve the physical meaning (although averaged) of the model parameters. Wigmosta and Prasad (2006) define a model as scalable when process equations, and model assumptions remain valid under upscaling or downscaling of its variables. The parameters, however, may require reconsideration to encompass the influences of spatial or temporal variability encountered at the extrapolated scale.

Among the various statistical downscaling approaches, empirical downscaling methods are the most commonly used due to their ease of implementation. These empirical methods are often grouped into change factor and bias correction methods, depending on their use of calibration strategies (Chen et al., 2013). These methods can be considered as an additional post-processing step applied to low resolution model outputs.

Statistical downscaling of a hydrological variable (e.g. soil moisture) can be also achieved by using topographic attributes or indices (e.g. slope, wetness index) to explain its spatial variability. Topographic indices have been widely applied in hydrology (Beven and Kirkby, 1979; Moore et al., 1991). These assume that topography is dominant in controlling and modifying the soil hydrologic processes operating in the landscape (Western et al., 1999; Savenije, 2010). Western et al. (1999) analyse the predictive ability of several terrain indices against soil moisture data collected in a small humid catchment in south-eastern Australia. During wet conditions the logarithm of specific area and wetness index explained, respectively, 50% and 42% of the spatial soil moisture variance. They noticed that as the catchment dried out, the explanatory power of the indices dropped off rapidly (Blöschl, 2006). Blöschl et al. (2009) develop a downscaling method for scatterometer soil moisture data based on hydrological concepts. They take topography, soils and climate as the main controls into account.

In this chapter, we explore the statistical downscaling of a physically based model in which the process equations are assumed invariant of the spatial scale but the input parameters maps are specific for two different resolutions (one is 100 times larger than the other). We base this analysis on a "high" resolution- hydrological model for the Limpopo river basin (0.05° x 0.05°) and the original "low"-resolution hydrological model for entire Africa (0.5° x 0.5°). We use the results of the two models to compute the effect of spatial resolution on three distributed fluxes or storages: actual evaporation, soil moisture, and total runoff. We assess the potential for downscaling of hydrological model outputs with a simple bias correction approach and a more comprehensive approach based on physical characteristics of the basin.

5.2 Materials and methods

5.2.1 Study area and data

5.2.1.1 The Limpopo River basin

The Limpopo River basin is described in Section 4.2.1 and its location and the location of selected hydrometric stations for this study is presented in Figure 4-1.

5.2.1.2 Data used for model set up and forcing

The Digital Elevation Model (DEM) that is used in the basin is based on the Hydro1k Africa (USGS EROS, 2006). The majority of the parameters (maps) required for the models (e.g. soil layer depths, soil storage capacity, hydraulic conductivity etc.) were derived from three maps and their properties: Digital Soil Map of the World (FAO, 2003), the distribution of vegetation types from GLCC (USGS EROS, 2002; Hagemann, 2002), and the lithological map of the world (Dürr et al., 2005). Further details can be found in Chapter 4 and in Trambauer et al. (2014b). The parameters were computed for the two different resolutions by carefully interpolating and aggregating these datasets. Areas in the basin that are irrigated were obtained from the global map of irrigated areas in 5 arc-minutes resolution based on Siebert et al. (2007) and FAO (1997). We compute the monthly irrigation intensities per grid cell using the irrigated area map, the irrigation water requirement data per riparian country of the basin (DEWFORA, 2012a) and the irrigation cropping pattern zones (FAO, 1997).

All meteorological forcing data (daily precipitation, daily minimum and maximum temperature at 2 meter) that is used are based on the ERA-Interim (ERAI) reanalysis dataset from the European Centre for Medium-Range Weather Forecasts (ECMWF) (see Sections 3.2.1.1 and 4.2.2). Runoff data were obtained from the Global Runoff Data Centre (GRDC; http://grdc.bafg.de/), the Water Affairs Republic of South Africa and ARA-Sul (Administração Regional de Águas, Mozambique).

5.2.2 Setting up low and high resolution hydrological models

A process based distributed hydrological (water balance) model based on PCR-GLOBWB (van Beek and Bierkens, 2009) is used. This model is described in Sections 3.2.1.1 and 4.3.1. The PCR-GLOBWB global hydrological model was first applied at continental scale across Africa (Chapter 3, Trambauer et al., 2014a). The continental model, which has a spatial resolution of 0.5° x 0.5°, was cut out only for Limpopo river basin with the same spatial resolution. This we call the "low-resolution (L)" model hereafter. Then, we developed another model for the same basin using the same PCR-GLOBWB codes but with all the input parameters derived at a resolution of 0.05° (approximately 5 km). This we call the "high-resolution (H)". Both models have a daily temporal resolution. Both models were run for the period January 1979- December 2010 with the input parameter and meteorological data described in section 5.2.1.2. In the current versions of the model no calibration is applied to maintain the physical meaning of the model parameters (Trambauer et al., 2014b).

The two models were verified against runoff observations for the station at Chókwe (station no. 24, see Figure 4-1), which is the station with the largest contributing area with available data. We evaluate quantitative statistics as proposed by Moriasi et al. (2007), which comprise the Nash-Sutcliffe efficiency (NSE), and the ratio of the root mean square error to the standard deviation of the measured data (RSR). The coefficient of determination (R^2) is also included. These results are presented in Table 5-1. Based on these three statistics both models can be seen to perform reasonably well (NSE > 0.5, RSR ≤ 0.70, R^2 > 0.7), though the high resolution model does perform better. As a result, the high resolution model is used as the reference for the bias correction method.

Table 5-1 Model evaluation measures for runoff at Chókwe station for the two models with high and low resolution.

Evaluation measure	High res. model	Low res. model
R²	0.92	0.73
NSE	0.90	0.66
RSR	0.32	0.58

5.2.3 Comparison of variability of hydrological fluxes from the high and low resolution models

To have a better understanding of the two resolution models results, we assess the flux variability before we undertake the downscaling. We compare the fluxes or storages from two hydrological models, the "high" resolution model (0.05° x 0.05°) and the "low" resolution model (0.5° x 0.5°) (see Figure 5-1). The forcing precipitation and temperature data are unchanged for both resolution models. From both hydrological models resolutions (L: low and H: high), we analyse the results of three distributed fluxes or storages: actual evaporation (AET, mm month^{-1}), soil moisture (SM, mm month^{-1}), and total runoff (q_{tot}, mm month^{-1}). The total runoff is the sum of the different runoff components: direct runoff (q_1, mm month^{-1}), interflow or subsurface flow (q_2, mm month^{-1}), and baseflow (q_3, mm month^{-1}), and the soil moisture is the total storage in the two soil layers. We used the coefficient of variation ($CV = \frac{\sigma}{\mu}$, defined as the ratio of the standard deviation σ to the mean μ) to assess the variability of the high resolution variables within each low resolution pixel.

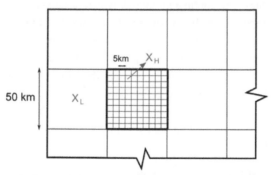

Figure 5-1 Grid representation of the low (L) and high (H) resolution models.

5.2.4 Downscaling of hydrological fluxes using bias correction approach

Empirical downscaling methods are the most commonly used amongst the various statistical downscaling approaches due to their ease of implementation. These use the observed or modelled variables of interest as predictors (Jakob Themeßl et al., 2011; Wang et al., 2015). Empirical methods are often grouped into change factor and bias correction methods, depending on the calibration method used (Chen et al., 2013). Bias correction methods assume that the discrepancies between observation and model simulations stay constant in time, while change factor based methods assume that the change from present day to future in the observed climatology will be the same as the change in the modelled climatology. The downscaling methods in each group are classified into five types (i.e. mean based (MB), variance based (VB), quantile mapping (QM), quantile correcting (QC) and transfer function (TF)), depending on the statistical transformation used (Chen et al., 2013; Wang et al., 2015).

We evaluate two bias correction methods, a variance based method and a quantile mapping method (see Table 5-2). These can be interpreted as post-processing applied to the low resolution model output to attain the higher resolution variables. In this procedure, the high resolution model outputs are considered as the reference. These methods assume that the discrepancies between the high model results (X_H) and the low model results (X_L) stay constant in time. We use monthly maps for a 32 year period (1979 - 2010, 384 months) of the distributed variables. Of these we use 20 years (240 months) for calibration and 12 years (144 months for validation). We evaluate the same hydrological variables as described above (AET, SM, and q$_{tot}$).

Table 5-2 Empirical statistical downscaling methods. Subscript 'H' represent high resolution and 'L' represents low resolution; with or without apostrophe represents validation and calibration period respectively; F and F^{-1} represent Empirical distribution function (ECDF) and reverse ECDF, respectively. Table is based on Wang et al. (2015).

Method	Description	Transformation
Variance based VB	Corrects both the mean and variance of X_L to match those of X_H	$X'_H = \frac{(X'_L - \mu_L)}{\sigma_L} \cdot \sigma_H + \mu_H$
Quantile mapping QM	A value of X_L, e.g. with the percentile p in the X_L distribution is corrected as the value with the same percentile in the X_H distribution.	$X'_H = F_H^{-1}[F_L(X'_L)]$

For each method, we apply an independent transformation for each month of the year. To evaluate the model error predictions we use the Nash-Sutcliffe efficiency (NSE) and the normalised root mean squared error (NRMSE), defined as the RMSE divided by the sample mean. The two evaluation measures are computed separately for each month of the year and for each high resolution pixel, resulting in 12 spatially distributed maps for each evaluation measure.

Additionally, we assess the performance of the downscaling methods for the validation period on the monthly time-series using the modified Kling-Gupta Efficiency (KGE) (Kling et al., 2012; Thiemig et al., 2013). The KGE includes three components: correlation (r), bias (β), and a measure of variability (γ) and is defined as follows:

$$KGE = 1 - \sqrt{(r - 1)^2 + (\beta - 1)^2 + (\gamma - 1)^2} \tag{5-1}$$

$$r = \frac{\sum_{i=1}^{n}(X_{DS,i} - \overline{X_{DS}})(X_{H',i} - \overline{X_{H'}})}{\sqrt{\sum_{i=1}^{n}(X_{DS,i} - \overline{X_{DS}})^2}\sqrt{\sum_{i=1}^{n}(X_{H',i} - \overline{X_{H'}})^2}} \tag{5-2}$$

$$\beta = \frac{\mu_{DS}}{\mu_{H'}} \tag{5-3}$$

$$\gamma = \frac{CV_{DS}}{CV_{H'}} = \frac{\sigma_{DS}/\mu_{DS}}{\sigma_{H'}/\mu_{H'}} \tag{5-4}$$

Where r is the Pearson correlation coefficient, μ is the mean value of the variable, CV is the coefficient of variation, σ is the variable standard deviation, and the indices H' and DS represent the high resolution for the validation period and downscaled variables. KGE, r, β, and γ are dimensionless with perfect efficiency values at one. Thiemig et al. (2013) classify a model performance based on KGE as follows: good (KGE ≥ 0.75), intermediate (0.75 > KGE ≥ 0.5), weak (0.5 > KGE > 0.0) and poor (KGE ≤ 0.0). For the KGE we have only one distributed map indicating the performance during the whole validation period without distinction of season or month.

5.2.5 Downscaling of hydrological fluxes using landscape characteristics

Here we define a downscaling method that uses topographic characteristics (Eq. 5-5). The idea is to investigate if the results from the low resolution (L, continental scale) hydrological model can be transferred to the high resolution (H, catchment scale) model by identifying spatial patterns based only on topographical features. The Height Above the Nearest Drainage (HAND, Rennó et al., 2008) and slope have been identified as a powerful tool to be used for hydrological classification (Savenije, 2010; Gharari et al., 2011). Hence, the topographical characteristics used as explanatory variables are the HAND and the slope. The HAND was computed using the procedure described in Rennó et al. (2008) and Nobre et al. (2011) for the high resolution model. For each variable, we compute the high resolution X_H (mm month^{-1}) as a function of the spatial pattern of the HAND (m), the slope (%) at the high resolution, and the low resolution variable X_L (mm month^{-1}):

$$X_H = f(HAND, slope, X_L)$$ (5-5)

The functions evaluated (represented with f in Eq. 5-5) are the multiple linear regression (MLR) and decision tree (DT) models. The MLR model establishes a linear transfer function between one or more predictors and the predictand such that

$$X_H = \alpha + \sum_{p=1}^{l} \beta_p X_p + \varepsilon$$ (5-6)

with α being the intercept or independent term, β the regression coefficients, ε the error, X_p the p predictor variables and X_H the predictand (Jakob Themeßl et al., 2011). The decision tree (regression) is a machine-learning method for constructing prediction models from data. Prediction models are obtained by recursively partitioning the data space and fitting a simple prediction model within each partition. As a result, the partitioning can be represented graphically as a decision tree. In decision trees, the dependent variables can either be a continuous or ordered discrete values (Loh, 2011).

The calibration/validation sets were divided randomly, leaving 70% of the data for calibrating the models and 30% for validation. The performance of both the MLR and DT downscaling methods for the validation dataset is measured by using the linear correlation and the modified Kling-Gupta Efficiency (KGE) between the observed and predicted values.

From an exploration of other variables, it seems that the soil characteristics also influence the spatial variability of the high resolution variables. To illustrate this, Figure 5-2 presents the low (X_L) and high (X_H) resolution total runoff (q_{tot}) for March 1996. Some of the patterns of the $q_{tot,H}$ seem to follow the soil characteristics in the basin (see Figure 5-3). Figure 5-4 presents the long term average of the different runoff components with the soils or lithology differentiation superposed. It is clear that the direct and subsurface runoff patterns closely follow the soil differentiation in the basin, and the baseflow follows the lithology differentiation. For instance, highest direct runoff occurs in soils with the highest residual volumetric moisture content in the top layer. To illustrate this we show two areas (A and B) in Figure 5-4 (top panel) corresponding to soil classes in which the direct runoff is higher than in surrounding areas. These same areas can be visualized in Figure 5-3. Similarly, highest subsurface flow occurs for intermediate residual volumetric moisture content in the top layer (see for example areas C and D in Figure 5-3 and Figure 5-4). The residual moisture influences the soil's infiltration capacity. The higher the antecedent soil moisture resulting from the higher residual moisture content, the more quickly the soil becomes saturated, initiating runoff. Lower residual volumetric moisture content allows water to infiltrate to the second layer contributing to q_2. Highest baseflow occurs for highest recession coefficient for the groundwater storage (see areas E and F in Figure 5-3 and Figure 5-4). Moreover, there is a region in the centre of the basin, where the patterns of direct runoff and subsurface flow seem to be explained by either the fraction of tall vegetation or by the HAND. For this reason, to evaluate this we decided to include all the parameters presented in Figure 5-3 as possible explanatory variables in our downscaling models. Thus, we calculate the high resolution variable, but also considering parameters directly related to the soil, lithology and vegetation:

$$X_H = f(HAND, \ slope, \ soil, \ lith, \ veg_{tall}, \ X_L) \tag{5-7}$$

where soil is represented by one of its properties such as the residual volumetric moisture content in the top layer (m³.m⁻³), lithology (lith) is represented by the recession coefficient for the groundwater storage (day⁻¹) and veg$_{tall}$ (-) is the fraction of tall vegetation within each grid cell. The functions evaluated (f) are once again the multiple linear regression (MLR) and decision tree (DT) models.

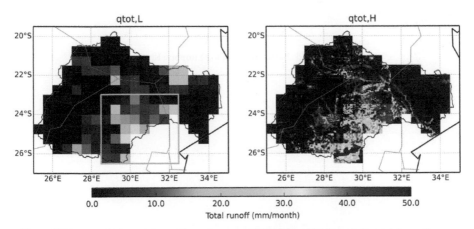

Figure 5-2 Low resolution total runoff q$_{tot,L}$ (mm month⁻¹) (left), and high resolution total runoff, q$_{tot,H}$ (mm month⁻¹) (right) for March 1996. The green square highlights a selected area with higher flows.

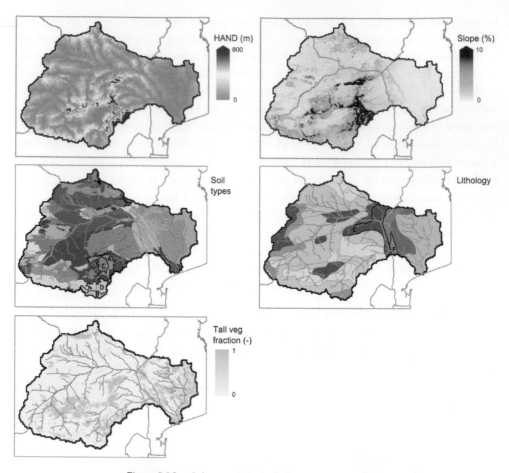

Figure 5-3 Land characteristics in the Limpopo River basin.

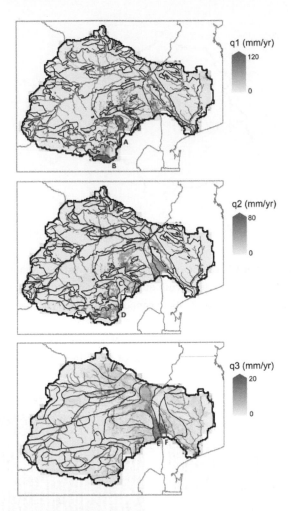

Figure 5-4 Long term average of runoff components with soil differentiation (top two panels) or lithology differentiation (lower panel).

5.3 Results and discussion

5.3.1 Variability of the high resolution fluxes within each low resolution grid cell

The time series of maps of the coefficient of variation (CV) of model output fluxes or storages can be visualised in an animation. We summarise the results by presenting Hovmoller diagrams (time-latitude and time-longitude plots) for selected longitude and latitude bands (see Figure 5-5).

For the actual evaporation, low coefficients of variation (CV below 1.5) can be observed for each latitude/longitude band (see Figure 5-6). The highest CV generally occurs in the winter (dry period). Spatially, there seems to be some indication that higher CV occur in the western part of the basin (drier part) and lower CV occur in the eastern part of the basin (wetter part).

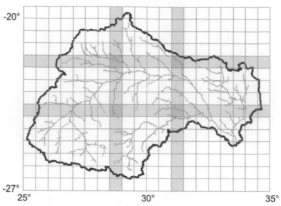

Figure 5-5 Chosen latitude and longitude bands for the Hovmoller diagrams. The grid shown is that of the low resolution model.

Figure 5-7 shows that higher CV can be observed for soil moisture. Again, similar to the actual evaporation flux, highest CV generally occurs during the winter (dry period) and the lowest CV occurs during the summer (wet period). Wet years can also clearly be seen as periods with very low CV. In the same way, the pattern of the CV for total runoff shows lower CV for wet periods (Figure 5-8). However, there is a clear change in the pattern after the flood of 1999/2000. To have a better understanding of this change, the variability of the different runoff components was also analysed.

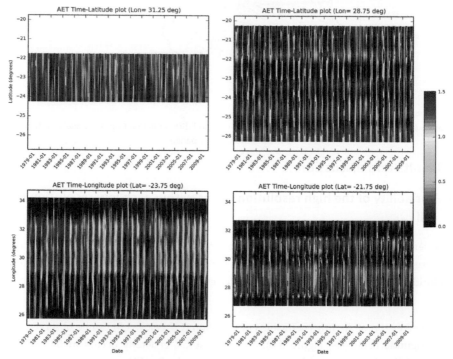

Figure 5-6 Latitude-time plots (top) and Longitude-time plots (bottom) of the grid (0.5°) monthly coefficient of variation of actual evaporation results from the high resolution model.

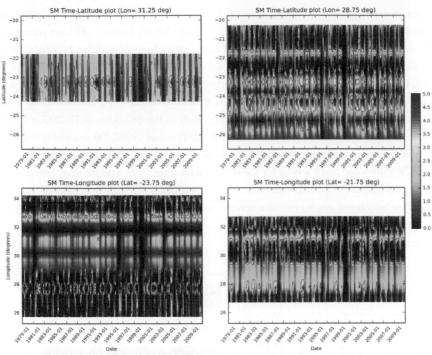

Figure 5-7 Latitude-time plots (top) and Longitude-time plots (bottom) of the grid (0.5°) monthly coefficient of variation of soil moisture results from the high resolution model.

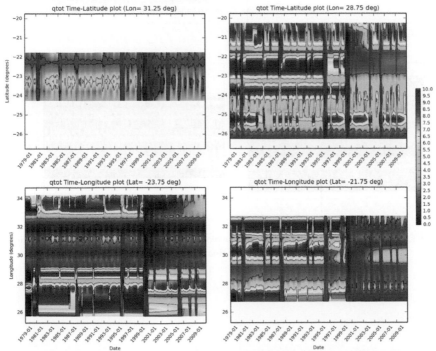

Figure 5-8 Latitude-time plots (top) and Longitude-time plots (bottom) of the grid (0.5°) monthly coefficient of variation of total runoff results from the high resolution model.

For the runoff contributions, the CV is found to be much higher than for actual evaporation and soil moisture, as can be seen in the range shown in Figure 5-8. In the case of direct runoff (Figure 5-9) and subsurface flow (not shown), horizontal bands of higher CV (red bands) can be observed in the plots, implying that particular cells have higher CV than others for most of the time steps. In general, these cells are those with higher direct runoff values. Another interesting detail when analysing these horizontal bands of higher CV is that anomalous dry periods are identified by longer periods with higher CV (longer red periods) such as the droughts of 1982/83 and 1991/92. In contrast, wet periods such as 1980/81, 1995/96 or the flood of 1999/2000 present low CV values which can be observed as clear interruptions of these lines. With respect to baseflow (Figure 5-10) very high CV values can be observed for the entire basin until 1999. The flood of 1999/2000 had a significant contribution to groundwater recharge, affecting the hydrological system, and the CV of the baseflow remained much lower for several years in some areas. From these plots, it is clear that the change that occurred in the CV after the flood of 1999/2000 in the total runoff was due to the change in CV in the baseflow.

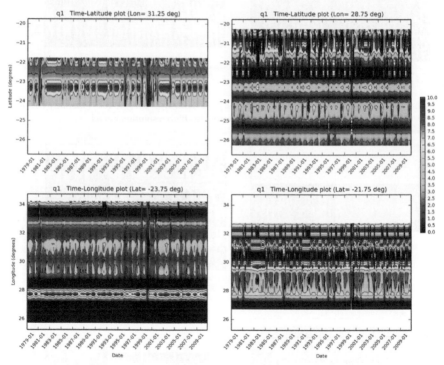

Figure 5-9 Latitude-time plots (top) and Longitude-time plots (bottom) of the grid (0.5°) monthly coefficient of variation of direct runoff results from the high resolution model.

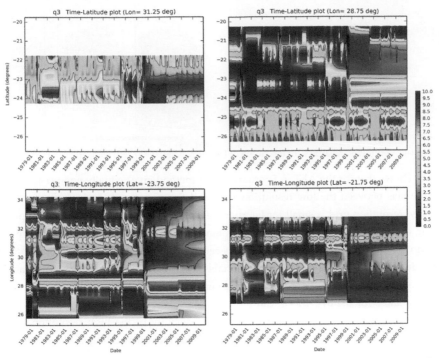

Figure 5-10 Latitude-time plots (top) and Longitude-time plots (bottom) of the grid (0.5°) monthly coefficient of variation of baseflow results from the high resolution model.

5.3.2 Downscaling based on bias correction

For each distributed model variable we calibrate using both bias-correction methods presented in Table 5-2 with 20 years of monthly time series (1979-1998, 240 months) and we validate with 12 years (1999-2010, 144 months). This results in a transformation for each month of the year and for each of the two methods. As described in Section 5.2.4 we used the NSE, NRMSE, and KGE as validation measures. Initially, we also analyse mean based (MB, additive and multiplicative) methods, but given that the performance of the VB and QM methods was slightly better, and knowing that these two methods are more complete than the mean based methods, we centre our analysis on the VB and QM methods.

Figure 5-11 present the evaluation measures NSE (left) and NRMSE (right) for the validation of the different variables and different downscaling methods for the month of February. It is clear from the figures that for both methods evaporation (AET) and soil moisture (SM) can be downscaled with higher efficiency or lower error than the total runoff. The quantile mapping method gives more continuous distributions than the variance based method. For total runoff and SM (for NRMSE) we can see that the efficiency varies between the methods used, but some clear zones or areas with higher efficiency than others are noticed. Large areas without data (white areas) in q_{tot} result from the high amount of zero values in the flows, due to the aridity of the area. The downscaling methods for drier months and areas generally result in lower efficiencies, which could be expected from previous studies (e.g. Western et al., 1999).

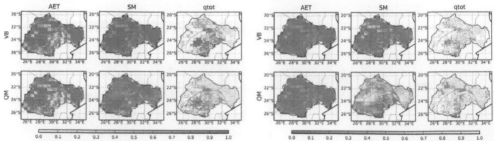

Figure 5-11 Verification measures for the month of February (validation), left: NSE, and right: NRMSE.

Figure 5-11 shows that for the AET in February either method gives equally good results. This is because the bias between the low resolution AET and the average of the high resolution AET over the low resolution grid is small for February. The highest bias between these two products concerning evaporation occurs from May to September. During these months the downscaling methods do not produce such good results. This can be seen in Figure 5-12 where we present a time-latitude plot of the NSE for the three variables (AET, SM, and q_{tot}) for the VB and QM methods. For SM a similar pattern is observed; a very good performance for the wet season and poor performances for the dry season can be recognised. These results could be expected from the results in the previous section, that the higher variability of the high resolution variables during the dry (winter) months makes it more difficult for the downscaling model to capture the variability.

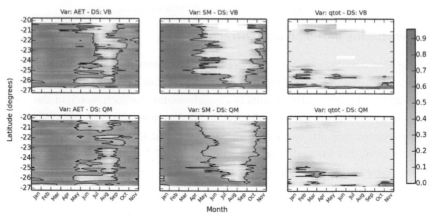

Figure 5-12 Time-Latitude plots (mean value over Longitudes) of NSE for AET (left), SM (middle) and q_{tot} (right) for both bias correction methods used; VB (top) and QM (bottom).

When SM is compared to AET, we see higher efficiencies (higher NSE), but also higher errors (higher NRMSE) in the month of February (see Figure 5-11). This is probably because in the computation of the NSE the bias is normalized by the standard deviation of the high resolution SM (SM_H). As a result of the higher variability in the SM_H compared to that in the AET_H, the bias has a lower 'weight' in the computation of NSE, resulting in higher efficiencies. For total runoff we observe some areas that perform well, but the performance in general is low when compared to AET and soil moisture. Figure 5-13 presents the Kling-Gupta Efficiency (KGE) for the VB and QM methods during the whole validation period. Good model performances are observed for

AET and SM. For total runoff good or intermediate performances are found in some areas, but also poor performances are observed in large parts of the basin. Figure 5-13 shows again that the QM method gives more continuous distributions than the VB method. These more continuous patterns seem more realistic.

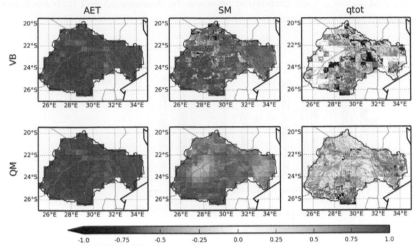

Figure 5-13 Kling-Gupta Efficiency (KGE) (validation) for variance based (VB) and quantile mapping (QM) methods. Model performance: good (KGE ≥ 0.75), intermediate (0.75 > KGE ≥ 0.5), weak (0.5 > KGE > 0.0), and poor (KGE ≤ 0.0) (Thiemig et al., 2013).

5.3.3 Downscaling based on landscape types

To analyse the downscaling of the low resolution hydrological model results to high resolution based only on the terrain characteristics (only HAND and slope as predictors) we evaluated a multiple linear regression model (MLR) and a decision tree approach as explained in Section 5.2.5, Eq. (5-5). The calibration/validation sets were divided randomly, leaving 70% of the data for fitting the model and 30% for validation. The results of a particular wet month, March 1996, for the total runoff variable (q_{tot}) indicate that a simple multiple linear regression model can explain 43% of the spatial variability for the validation set. Even though these results are particular to the calendar month, there is some indication that the terrain characteristics can help to explain part of the direct runoff variability.

When adding other explanatory variables as explained in Eq. (5-7) the results improved. For example, for March 1996 the multiple linear regression model for total runoff for the whole basin could explain 49% of the spatial variability. For more sophisticated tools the performance increases. When a decision tree scheme is used the model explains 66% of the spatial q_{tot} variance. Given the high aridity and low flows in parts of the basin, we narrowed the model boundaries to the area where the flows are not near zero. This area is highlighted in Figure 5-2 (left). The results of the MLR explain 52% of the spatial variance of q_{tot}, while the decision tree approach explains 77% of the spatial variance.

Figure 5-14 presents the KGE and its three components for the downscaled models (MLR and DT) for total runoff for the month of March 1996 for the entire basin and for the selected area.

Based on the criteria defined in Section 5.2.4, the performance of the MLR model is intermediate both for the entire basis and the selected area, and the performance of the decision tree is good for both the entire basin and the selected area. Figure 5-15 presents a scatter plot of the high resolution total runoff and the downscaled total runoff with the decision tree model for the selected area, and shows a good correlation between the downscaled and high resolution total runoff (correlation coefficient $r = 0.88$).

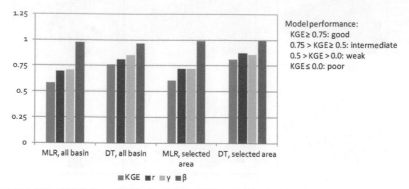

Figure 5-14 Modified Kling-Gupta efficiency (KGE) for total runoff (qtot) for the downscaling models: Multiple Linear Regression (MLR) and Decision Tree (DT), for March 1996.

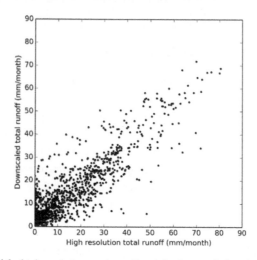

Figure 5-15 Scatter plot of the high resolution total runoff and the downscaled total runoff for March 1996 with the decision tree model for the selected area.

The results just presented correspond to a particular month or time step. A similar analysis was performed by Western et al. (1999) and Green and Erskine (2004), who evaluate the performance of terrain indices in reproducing the spatial pattern of soil moisture for particular sampling campaigns. This allows evaluating the performance of the downscaling model for that month, but requires both resolution models for this evaluation. We are interested in finding a downscaling model based on land characteristics that would allow transferring the low resolution information to a higher resolution for any month without having the high resolution map beforehand. As conditions change throughout the year, a different downscaling model is developed for each month. If we have a calibrated downscaling model for each month of the year, we can then use the model as an estimating tool for the high resolution flow of a given month whenever the low resolution flow of that month is available. With this aim, we fit 12 MLR models for each variable, one for each month. The performances of these models (for the selected area) for the validation period based on the KGE are presented in Figure 5-16.

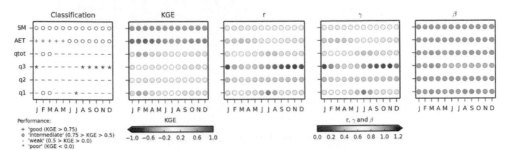

Figure 5-16 Performance of the MLR downscaling models based on land characteristics for the selected area.

The results show that the models perform quite good or intermediate for SM and AET. Similarly to other studies (e.g. Western et al., 1999), better performance is found in the wet months than in the dry months. These results are similar to those found for the bias correction method and presented in Figures 5-11, 5-12 and 5-13. For the runoff components, the models perform quite weak, but for a couple of months, February and March, the downscaling models can predict direct and total runoff quite well. Total runoff was also better predicted for the months of February and March in the bias correction method (see Figure 5-12). Baseflow is badly predicted for half of the year, from August to January, which corresponds to the months with lower baseflow in the basin. While the main rainy season occurs from October to March, the main runoff season occurs from December to May (Trambauer et al., 2015), and the highest baseflow occur from February to July. Baseflow is computed on a cell-by-cell basis and ignores the lateral exchange. From that perspective, the variability on a 0.05° grid cell may overestimate the variability and negatively affect the downscaling potential. From the different components of the KGE we can indicate that the bias ratios between the observed and predicted variables are in every case close to 1, which indicates a good agreement in the mass balance. The low γ values for the different runoff components indicate a poor representation of the flow variability.

The empirical statistical downscaling methods present both advantages and disadvantages. One of the main drawbacks of these methods corresponds to the temporal consistency on the models. To analyse this issue, we plot the regression coefficients resulting from the monthly MLR downscaling models based on the land characteristics for the selected area (see Figure 5-17). The figure shows that the multiple regression coefficients change quite smoothly throughout the year, indicating a smooth change in the models.

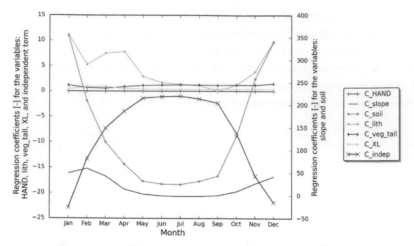

Figure 5-17 Regression coefficients for the monthly MLR downscaling models based on land characteristics for the selected area.

5.4 Conclusions

Global hydrological models usually have a spatial resolution or grid size in the order of 0.5 degrees, which is somehow limited by computational time and the spatial resolution of the global input data. However, this spatial resolution is not sufficient to adequately capture the spatial variability of processes in smaller basins. With the aim of recapturing the spatial variability of the processes that were smoothed out in the global models, it would be useful to downscale any of the resulting hydrological variables (e.g. evaporation, soil moisture, runoff, etc.) from these global hydrological models to finer spatial scales for any desired region.

First, the results from two hydrological models of the Limpopo River basin based on PCR-GLOBWB but with different spatial resolutions (0.5° x 0.5° and 0.05° x 0.05°) are compared to assess the scaling effect on the model results. Initially, using the results of the two models, we compute the effect of spatial resolutions on three distributed fluxes or storages: actual evaporation, soil moisture, and total runoff. We use the coefficient of variation (CV) to assess the variability of the high resolution variables within each low resolution pixel. The results are presented in Hovmoller diagrams, which plot the variability of the high resolution fluxes within each low resolution pixel as a function of time. These diagrams allow differentiating dry (higher CV) and wet seasons (lower CV) in the basin. Wet years can clearly be seen as periods with very low CV. Actual evaporation presents the lowest CV, followed by SM, and total runoff presents the highest CV.

Second, two different downscaling techniques are applied to downscale hydrological variables; bias correction statistical downscaling, and downscaling by using topographic and soil attributes as explanatory variables. The results from the two downscaling methods are similar. Evaporation can be downscaled successfully with good performance in the wet season. Evaporation is mainly controlled by: (i) atmospheric demand (vapour pressure deficit), (ii) available energy, and (iii) soil moisture and vegetation conditions. In both resolution simulations atmospheric demand and available energy were the same (coming from ERA-Interim forcing), while only soil moisture evolution and vegetation conditions varied. Our results suggest that for evaporation, the main drivers are the large scale atmospheric conditions, especially for the wet season (energy-limited conditions). For the dry season (water-limited conditions), the soil moisture and vegetation conditions play a bigger role. However, this conclusion is limited by our modelling methodology that used the same atmospheric forcing in both the high and low resolution simulations. Generally, meteorological data is not easily obtainable over poorly gauged catchments. This means that the resolution of the forcing over these basins will normally be lower than the resolution of the hydrological model. With the methods developed, a hydrological model with a scale commensurate with that of the meteorological forcing can be applied, with additional variability due to landscape and soil characteristics being added through these methods.

Third, the analysis of the variability of the fluxes on high resolution grid cells under different land features and soils indicates that there is good potential of downscaling the low resolution hydrological model results to high resolution based only on the terrain and soil characteristics. The performance of the downscaled results is better during the wet season than during the dry season. For a particular wet month, March 1996, downscaling methods based on land characteristics could explain up to 52% (MLR) or 77% (DT) of the total runoff spatial variance for the validation set of a selected area in the basin. The downscaling models for the whole period (1979-2010), where a different downscaling model is developed for each month of the year, perform quite good or intermediate for SM and AET. Better performance is found in the wet months than in the dry months. For the runoff components, the models perform quite weakly, but for a couple of months during the wet season the downscaling models can predict direct and total runoff relatively well (r = 0.71, KGE = 0.58). Baseflow is badly predicted during the dry baseflow season.

The downscaling of low resolution hydrological variables needs to be analysed further, and this initial approach remains empirical and the results site specific. Together with the development of global modelling, research should move towards better understanding the downscaling of hydrological models and variables. For some purposes such as flood predictions, where an initial fast assessment of the situation is desired, downscaling of hydrological variables results would be much more efficient than running high resolution hydrological models due to the computation time gained. However, when there is no time or economical constraints, setting up a high resolution model is preferred over downscaling of low resolution results.

HYDROLOGICAL DROUGHT FORECASTING AND SKILL ASSESSMENT FOR THE LIMPOPO RIVER BASIN

This chapter addresses the seasonal prediction of hydrological drought in the Limpopo River basin by testing three proposed forecasting systems that can provide operational guidance to reservoir operators and water managers at the seasonal time scale. All three forecasting systems use a distributed hydrological model of the basin, but with different meteorological forcings, namely (i) a global atmospheric model forecast (ECMWF seasonal forecast system - S4), (ii) the commonly applied ensemble streamflow prediction approach (ESP) using resampled historical data, or (iii) a conditional ESP approach (ESPcond) that is conditional on the ENSO (El Niño-Southern Oscillation) signal. We determine the skill of the three systems in predicting streamflow and commonly used drought indicators. We also assess the skill in predicting indicators that are meaningful to local end users in the basin. FS_S4 shows moderate skill for all lead times (3, 4, and 5 months) and aggregation periods. FS_ESP also performs better than climatology for the shorter lead times, but with lower skill than FS_S4. FS_ESPcond shows intermediate skill compared to the other two forecasting systems, but it is more robust. The skill of FS_ESP and FS_ESPcond are found to decrease rapidly with increasing lead time when compared to FS_S4. The results show that both FS_S4 and FS_ESPcond have good potential for seasonal hydrological drought forecasting in the Limpopo River basin, which is encouraging in the context of providing better operational guidance to water users.

This chapter is based on:

Trambauer, P., Werner, M., Winsemius, H. C., Maskey, S., Dutra, E., and Uhlenbrook, S.: Hydrological drought forecasting and skill assessment for the Limpopo River basin, southern Africa, Hydrol. Earth Syst. Sci., 19, 1695-1711, doi:10.5194/hess-19-1695-2015, 2015.

6.1 Introduction

Climate change studies show evidence of an intensification of the global water cycle (Huntington, 2006; IPCC, 2007a; Hansen et al., 2012; Trenberth, 2012; Coumou and Rahmstorf, 2012), with extreme events including floods and droughts expected to become more frequent. The UNISDR (United Nations Office for Disaster Risk Reduction) Hyogo Framework of Action 2005-2015 (UNISDR, 2005) describes early warning systems and action plans triggered on the issuing of a warning as one of the most effective strategies to mitigate the impacts of natural hazards. Operational forecasting of streamflow to inform early warning is already commonplace in several parts of the world, but the main focus is often on flood prediction. Operational forecasting of streamflow for drought prediction has to date not been applied as widely, despite the widespread recognition of the relevance and importance of drought forecasting in the research community. However, in recent years several drought monitoring and early warning systems have been established around the world, which focus on different types of drought and on different regions. Some of these are still under development and remain experimental at this stage. A review on the existing drought early warnings systems is presented in Section 1.1.6.

Yuan et al. (2013) applied the NCEP's Climate Forecast System version 2 (CFSv2) combined with the Variable Infiltration Capacity (VIC) land surface model for seasonal drought prediction over Africa. They used both the standardized precipitation index (SPI) and soil moisture as indices and the Brier skill score (BSS) to assess the probabilistic drought hindcast for 1982-2002. Their results show relatively good skill in the dry season but only limited skill in the rainy season. They indicate that CFSv2 precipitation is correlated with the observed precipitation over southern Africa, but only accounts for 44-45% of the variance of observations. They point out that for two extreme droughts CFSv2 predicted neutral conditions or only a weak anomaly. Our study focuses on the Limpopo River basin and follows a similar type of analysis, although it does so at a higher resolution and in a more detailed manner. Additionally, we present different skill scores for different hydrological drought indicators during the rainy season, and compare different forecasting systems in the basin. We focus on assessing the skill of the forecast in predicting indicators that are meaningful to the local end users in the basin.

The semi-arid Limpopo River basin, located in southern Africa, has experienced severe droughts in the past, which have led to crop failures, high economic losses and the need for humanitarian aid. An effective drought early warning system for this basin is of prime importance. Current practices for drought forecasting in the Limpopo River basin involve three forms of seasonal climate forecasts ranging from regional to local scales: The Southern Africa Regional Climate Outlook Forum (SARCOF) climate outlooks, seasonal climate outlooks prepared by meteorological departments, and forecasts based on local knowledge applied in rural communities. Despite these seasonal forecasts being available in the basin, farmers seem to prefer to rely on drought forecasting systems based on indigenous and traditional knowledge. Such forecasts include signs in (i) the sun, moon and wind; (ii) trees and plants; and (iii) insects, birds and animals (DEWFORA, 2013b). For seasonal forecasts to be accepted by the local community there are several challenges that need to be addressed. End users should receive the information in a suitably understandable format at the time they need it for the forecast to be useful. The highly technical information that is typically contained in the forecasts should then be translated to a comprehensible form before being disseminated and delivered to decision makers and farmers. Moreover, end users should be involved in the product verification by providing feedback to the forecasters (DEWFORA, 2012b).

Seasonal hydrological drought forecasts aim for high hydrological predictability at a seasonal timescale. Shukla et al. (2013) quantified the contribution of a good representation of initial hydrologic conditions (IHCs) and seasonal meteorological forecast (MF) to seasonal hydrological predictability at different forecast dates and lead times (1, 3, and 6 months) globally. They quantified the contributions of two components of the IHCs (soil moisture and snow water content) through ensemble streamflow prediction (ESP) and reverse-ESP. Their results show that for the region of the Limpopo River basin the MF dominates the hydrological predictability during the wet season (forecasts starting in October and January) for almost every lead time considered. Only for the 1-month lead time forecasts issued in October did the IHCs appear to some extent to have a higher influence. For the dry season the IHCs dominate the hydrological forecast at all lead times. These results suggest that to improve the seasonal hydrologic forecast skill in the Limpopo River basin, efforts should focus on improving the MF. However, the contribution of other IHCs (surface water and groundwater level) to hydrological predictability should also be assessed.

Yossef et al. (2013) also investigated the relative contribution of initial conditions and meteorological forcing to the skill of the global seasonal streamflow forecasting system FEWS-World, using the global hydrological model PCR-GLOBWB (PCRaster Global Water Balance). They use ESP and reverse-ESP to determine the critical lead time for different locations at which the importance of the initial conditions is surpassed by that of the meteorological forcing. They indicate that for semi-arid regions such as the Limpopo Basin the initial conditions do not contribute much to the skill given the high sensitivity of the runoff coefficient to rainfall variability. This would suggest that the predictability in semi-arid basins such as the Limpopo using ESP is limited, with seasonal meteorological forecasts potentially offering better skill.

In this study we introduce three dynamic forecasting systems based on a distributed hydrological model for the seasonal prediction of hydrological droughts for the semi-arid Limpopo Basin in southern Africa. All three forecasting systems include a distributed hydrological model of the basin, and are forced by either (i) a global atmospheric model (ECMWF seasonal forecast system S4), (ii) the ESP approach using resampled historical data, or (iii) a conditional ESP approach (ESPcond) that is conditioned on the ENSO (El Niño-Southern Oscillation) signal. The aim of this study is to assess the skill of the three systems in predicting meaningful drought indicators for the Limpopo Basin.

6.2 Methods and data

The approach followed in this study is summarized in Figure 6-1. It starts with obtaining the meteorological seasonal forecast and preprocessing the data. This is then used to force the hydrological model (embedded in the Delft-FEWS forecasting shell (Werner et al., 2013)), thus obtaining seasonal forecasts of streamflow and other hydrological variables.

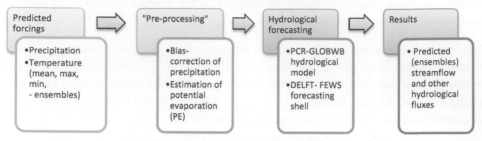

Figure 6-1 Approach followed in this forecasting system for the Limpopo River basin.

6.2.1 Ensemble hydrological forecasting in the Limpopo river basin

6.2.1.1 Study area - Limpopo river basin

The Limpopo River basin is described in Section 4.2.1 and its location is presented in Figure 4-1. Rainfall in the basin is seasonal (mainly from October - April), influenced by the movement of the intertropical convergence zone. Moreover, rainfall is highly variable causing frequent droughts, though floods can also occur during the rainy season. In the period 1980-2000, the southern African region was stuck by four major droughts in the seasons 1982/83, 1986/87, 1991/92 and 1994/95. This corresponds to an average frequency of a drought every 4 or 5 years, although the periodicity of droughts is not necessarily predictable. It is estimated that during the 1991/92 drought in southern Africa 86 million people were affected, 20 million of whom were considered to be at serious risk of starvation (DEWFORA, 2011).

Hydrological modelling in the Limpopo Basin is extremely challenging due to its very low runoff coefficients (see Section 4.2.1). Even a small error in precipitation or evaporation estimates could result in quite a large error in runoff estimation. Moreover, the uncertainty in the rainfall input could easily be larger than the runoff coefficient (4.3%) of the basin. Figure 6-2 shows the location of selected runoff stations and reservoirs in the Limpopo Basin.

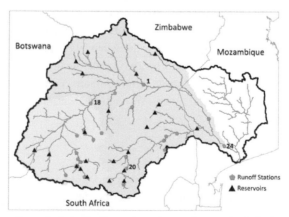

Figure 6-2 Locations of selected hydrometric stations and reservoirs in the Limpopo Basin.

6.2.1.2 The forecasting system

Regional hydrological model

A finer-resolution version ($0.05° \times 0.05°$) of the $0.5° \times 0.5°$ resolution global PCR-GLOBWB hydrological model (van Beek and Bierkens, 2009; van Beek, 2008) is used. This is a continuous-time simulation, process-based distributed model applied on a raster basis. The model is set up for the Limpopo Basin with a spatial resolution of $0.05° \times 0.05°$ and the simulation is carried out with a daily time step. A description of the model setup is presented in Sections 4.2.2 and 4.3.1. As the scope of this study is on the skill of the hydrological forecast, reservoirs are considered in a simple way. Cells with reservoirs in the model are considered as having a maximum storage volume. Releases to irrigation are taken into consideration as a fixed monthly value and subject to availability, and the reservoir will spill when full. The reservoirs in the basin are mainly used for irrigation.

Delft-FEWS shell

The hydrological model is embedded in the Delft-FEWS (Flood Early Warning System) open shell for forecasting purposes. The shell provides a sophisticated collection of modules designed for building a hydrological forecasting system customised to the specific requirements of an individual organisation. The philosophy is to provide an open shell for managing the data handling and forecasting process. This shell incorporates a comprehensive library of general data handling utilities, allowing a wide range of external models to be integrated in the system through a published open interface. This allows existing simulation models and data streams to be incorporated into a comprehensive and reliable forecasting system (Werner et al., 2013).

Reference run for the period 1979-2010

The hydrological model is run in simulation mode for a 32-year period (1979-2010) with the ERA-Interim forcing meteorological data at a daily time step. ERA-Interim (ERAI) is the latest global atmospheric reanalysis produced by the European Centre for Medium-Range Weather Forecasts (ECMWF) and is described in Sections 3.2.1.1 and 4.2.2. The ERA-Interim precipitation data was corrected using the GPCP (Global Precipitation Climatology Project) v2.1 product to reduce the bias when compared to measured products (Balsamo et al., 2010), as described in Section 3.2.1.1. This corrected version of precipitation was also used in the production of the ERA-Interim/Land data set (Balsamo et al., 2015).

In addition to the precipitation, other meteorological parameters from the ERA-Interim reanalysis data that are used to force the model include the 2m daily temperature (minimum, maximum and average). Temperature data is used for the computation of the reference potential evaporation that is required to force the hydrological model. In this study, the Hargreaves formula was used (see Section 4.2.2). The ERA-Interim data for the 32-year period from 1979 to 2010, corrected using the GPCP v2.1 data set are converted to the same spatial resolution as the continental-scale version of the PCR-GLOBWB model. ERAI is archived on an irregular grid (reduced Gaussian) with an approximate resolution of $0.7°$ over the domain. The data is downscaled from the ERAI grid to the original $0.5°$ model grid using bilinear interpolation and assumed to be constant over the 0.5° grid cell. No further downscaling of the meteorological forcings is carried out.

Initial conditions

The reference run provides the initial conditions for all forecasts. Initial conditions at the beginning of each month are saved in the Delft-FEWS database, and subsequently used as "warm states" to start the forecasts when doing the retroactive forecast (also referred to as hindcasts).

Time period of the simulations

An ensemble of meteorological hindcasts is first tested for the summer rainfall season over southern Africa for the period 1981–2010. Seasonal forecasts in this study are issued for only 7 months of the year so as to capture the rainy season and main runoff season (meaning the there are 5 months where we do not issue a forecast). The predictive skill for drought is expected to be higher during the dry season and lower during the wet season given that the hydro-climate has a longer persistence during the dry season (Yuan et al., 2013). Yuan et al. (2013) show the high contrast in skill between the dry and wet seasons in southern Africa.

In the hindcast, the first forecast of each season is issued in August and includes the seasonal (6 months) forecast from August to January. The forecast is updated at the beginning of each month from September to February. The last forecast of the season is issued in February, covering the period from February to July (see Figure 6-3). All simulations are done at a daily time step.

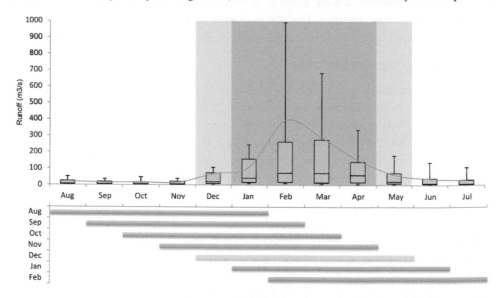

Figure 6-3 Upper plot: Limpopo River flow regime for Station 24 at Chókwe. The blue line represents the average observed runoff, and the whiskers of the boxplots represent the 10th percentile and the 90th percentile. The lighter and darker shaded areas represent the main runoff period and high runoff period, respectively. Lower plot: Initialization dates and length of forecasts during the year. The forecast issued in December is highlighted as the one that captures the main runoff season.

6.2.2 Seasonal forecasting systems

All three forecasting systems considered use the same hydrological model of the basin, but are forced with different meteorological forecasts. In the first system (FS_S4) the PCR-GLOBWB hydrological model is forced with the output of a global atmospheric model, the ECMWF seasonal forecast system S4 (atmosphere–ocean coupled). The second forecasting system (FS_ESP) is based on the ESP (Day, 1985) procedure. In the ESP procedure the ensemble meteorological forecast is generated with re-sampled historical meteorological data. The hydrological model is then forced with this re-sampled data. A third system (FS_ESPcond) is proposed given that the ENSO has a clear influence on the interannual climate variability over the Limpopo River basin (Landman and Mason, 1999). This is equivalent to the second system but the weights of the ESP ensemble members are conditioned on the ENSO signal (Oceanic Niño Index, ONI). This is explained in full in Section 6.2.2.3.

6.2.2.1 ECMWF S4 meteorological forecasts (FS_S4)

Meteorological ensemble forecasts

Seasonal meteorological forecasts from the most recent seasonal forecasting system at ECMWF (S4) are used to force the hydrological model. The S4 ensemble seasonal forecasts are initialised on the 1st of each month and the ensemble is generated by perturbations in the initial conditions and by the use of stochastic physics in the atmosphere during the model integration (out to 6 months lead time) (Molteni et al., 2011). The atmospheric resolution is about 79 km with 91 vertical levels, and is fully coupled with an ocean model with a horizontal resolution of 1°. S4 has been in operational use since November 2011, issuing 51 ensemble members with 6-month lead time. A hindcast set is provided for calibration and verification purposes, covering a period of 30 years (1981–2010) with the same configuration as the operational forecasts but with only 15 ensemble members. Molteni et al. (2011) presents an overview of S4 model biases and forecasts performance, and Dutra et al. (2013a, 2014) present an evaluation of S4 in seasonal forecasts of meteorological droughts. They found that S4-derived meteorological drought forecasts over southern Africa were skilful with up to 4-months lead time for SPI-6 in April. In the setup of FS_S4, the hydrological model is forced with the re-forecasts of the ECMWF seasonal system S4, with 15 ensemble members. A (hydrological) re-forecast is made to coincide with the 1st of each month in the 30-year hindcast set. Precipitation inputs to the hydrological model are accumulated from the 6-hourly S4 model values, while evaporation was calculated using the daily maximum and minimum temperatures directly archived by the meteorological model.

Climatological bias correction of seasonal forecasts of precipitation

Mean biases and drifts in the seasonal forecasts of precipitation can have a detrimental influence on the hydrological forecasts. Therefore, a simple climatological bias correction, based on monthly means, is applied to the seasonal forecasts in the form

$$P'_{m,l} = \alpha_{m,l} \, P_{m,l}, \tag{6-1}$$

where P and P' are the original and corrected seasonal forecasts of precipitation, respectively, α is a multiplicative correction factor and the subscripts m and l are the calendar month (1–12, of the initial forecast date) and lead time (0–5 months), respectively. The correction factor is given by the ratio

$$\alpha_{m,l} = \overline{P}_{m^*}^{base} / \overline{P}_{m,l} \, ,$$ (6-2)

where $\overline{P}_{m^*}^{base}$ is the climatological long-term mean of precipitation of the base data set for a particular calendar month m^* ($m^*=m+l$), and $\overline{P}_{m,l}$ is the long term ensemble mean of the forecasts for a particular month m and lead time l. The base data set used was ERA-Interim corrected with GPCP to be consistent with the baseline simulation period. The correction factor α is limited to a reasonable range (0.1–10), and is linearly interpolated from monthly values to daily values by assuming that it corresponds to day 15 of the particular month. Equation 6-1 is applied to the daily precipitation values. This is a simple bias correction that only guarantees that the mean forecast climate is similar to the climate of the base data set. It does not address other problems of the forecasts, common to all coupled atmosphere–ocean models, such as interannual variability, ensemble spread or daily variability.

6.2.2.2 ESP meteorological forecasts (FS_ESP)

A widely used approach to seasonal forecasting is the ESP procedure. ESP predicts future streamflow from the current initial conditions (warm state) in the hydrological model with re-sampled historical meteorological data (ERA-Interim-corrected with GPCP-observed meteorology from the last 31 years in this study). The procedure assumes that meteorological events that occurred in the past are representative of events that may occur in the future (Day, 1985). Although ESP is normally used in the absence of a seasonal forecast, in this study we use it to compare the skill of the FS_ESP with that of the FS_S4. Moreover, a comparison of these two forecasts may give an indication of what influences the predictability. ESP represents forecast uncertainty due to boundary forcing uncertainties only (Wood and Lettenmaier, 2008) and thus allows measuring the skill that can be expected only from initial states. In the FS_ESP hindcast, the sample of the year in which the forecast starts is excluded from the ensemble to allow for a fair estimate of the forecast uncertainty. The FS_ESP therefore includes 30 (31 minus 1) years in the ensemble.

6.2.2.3 Conditional ESP meteorological forecasts (FS_ESPcond)

ENSO is clearly related to interannual climate variability over the Limpopo River basin. In southern Africa meteorological droughts tend to happen in the December–March rainy season after onset of an El Niño event (Thomson et al., 2003). However, it is not always the case that this happens. Thomson et al. (2003) recorded a 120% increase in probability of drought disaster in the year after an El Niño onset. To account for the relationship between ENSO and the occurrence of drought, this system is similar to FS_ESP but the weights of ensemble members sampled through the ESP procedure are conditioned on the ENSO signal.

We use the post-ESP weighting technique described in Werner et al. (2004). This approach uses the El Niño - 3.4 index averaged over the 3-month-period immediately prior to the issue date of the forecast to weight ensemble members from ESP. The technique is summarised here for the forecast of the 6-month standardised runoff index (SRI-6).

1) Compute a vector (X) of absolute differences (x_i) between the value of the Niño - 3.4 index (ONI) in the forecast year and those of all the other years and sort the vector (X) from lowest to highest.

$$X = (x_1, x_2, \dots, x_n) \tag{6-3}$$

The sorted vector ($\aleph$) is

$$\aleph = \left[x_{(1)}, x_{(2)}, \dots, x_{(n)} \right], \ x_{(1)} \leq x_{(2)} \dots \leq x_{(n)} \tag{6-4}$$

2) Compute a vector of weights (W) for each member of the ESP ensemble by defining two parameters: a distance-sensitive weighting parameter (λ) and a parameter (α) that defines the k nearest neighbours used to calculate the weight of each member. Higher λ gives more weight to ensemble members with values of ONI closer to that of the forecast year. Higher α restricts attention to the n/α elements in the sorted vector. The ensemble member with the same year as the forecast year is assigned a weight of zero.

$$W = (w_1, w_2, \dots, w_n) \tag{6-5}$$

$$w_i = \left[1 - \frac{x_{(i)}}{x_{(k)}} \right]^{\lambda - 1}, \ x_{(i)} \leq x_{(k)} \tag{6-6}$$

$$w_i = 0, \ x_{(i)} > x_{(k)} \tag{6-7}$$

$$k = \text{NINT} \left(\frac{n}{\alpha} \right) \tag{6-8}$$

3) Calculate the probability (p_i) assigned to each ensemble member i by rescaling the weights.

$$p_i = \frac{w_i}{\sum_{j=1}^{n} w_j} \tag{6-9}$$

The parameters λ and α can be optimised for each case study or subbasin. The case with $\lambda = \alpha = 1$ is the traditional equal weighting scheme applied to ESP forecasts, with all ensemble members considered to have equal weight. If $\alpha = 1$ and λ varies, all ensemble members are considered, but these have non-zero weights that depend on the absolute distance between the ONI of the forecast year and the ONI of the year of the ensemble member. If $\lambda = 1$ and α varies only the nearest k ensemble members to the forecast year are considered in the ensemble, but they are all weighted equally. This case is similar to the approach applied by Hamlet and Lettenmaier (1999) for the Columbia River, where they restricted the ensemble members to those years that were similar in terms of the ENSO phase and the Pacific decadal oscillation phase. However, this restriction may result in ensembles with only few members, resulting in forecasts that are very sensitive to sampling errors (Brown et al., 2010). In the last case, where both α and λ vary, weights are assigned only to the k nearest ensemble members based on the distance of the index to the index of the forecast year (Werner et al., 2004). Werner et al. (2004) found this last case where both α and λ vary to show the best improvements for forecast skills.

For the FS_ESPcond we chose to keep the parameters constant ($\lambda = 2$ and $\alpha = 1$) given that the optimal selection of parameters would vary for each subbasin. Performing an in-depth selection of parameters for each subbasin is out of the scope of this study. Here we use $\lambda = 2$ and $\alpha = 1$, meaning that all ensemble members have a non-zero probability of being included in the ensemble, with that probability based on the distance between the ENSO indexes and the distance sensitive weighting parameter (linear for $\lambda = 2$). For each forecast start date, we construct an ensemble meteorological forecast of 30 members to be consistent with FS_ESP. The selection of the members is based on a resampling with replacement procedure given the probability assigned to each member. From the 30 possible ensemble members to be included, those with an ONI index closer to that of the forecast year, have a higher probability of being included in the ensemble. This means that some ensemble members are included more than once, and some are not included at all. The ONI indexes for the period 1979 -2010 were retrieved from NOAA (2014).

We also use this procedure for the forecast of SRI-4 (JFMA SRI). FS_ESPcond always uses the latest ONI index available prior to the start date of the forecast. This means that for the forecast issued in January, which corresponds to a 3-month lead time, FS_ESPcond uses the ONI values for October, November and December (OND). Similarly, for the forecast issued in December, which corresponds to a 4-month lead time, FS_ESPcond uses the SON (September, October, November) ONI, and the forecast issued in November (5-month lead time) makes use of the ASO (August, September, October) ONI.

6.2.3 Assessing skill of the forecasts

6.2.3.1 Skill scores

Standard verification skill scores are selected to measure the skill of the forecast ensembles in predicting drought indicators. In this study we use the Standardized Runoff Index (SRI) for the characterisation of hydrological droughts. This indicator is explained in Section 4.3.2.3. Forecasts are verified against the reference run and the resulting skills are established relative to sample climatology. Cloke and Pappenberger (2008) recommend the use of several verification measures in the same analysis so that the quality of the forecast can be assessed rigorously. We selected three verification scores that measure slightly different properties of the forecast skill. The ROC curve measures discrimination but not bias, the rank histogram measures reliability or bias, and the Brier score (BS) accounts both for reliability and sharpness (Renner et al., 2009).

The **ROC** (relative operating characteristic, or receiver operating characteristic) diagram measures the ability of the forecast to discriminate between two alternative outcomes. It plots the hit rate or probability of detection (POD) versus the false alarm rate or probability of false detection (POFD). It is not sensitive to bias in the forecast, so it says nothing about the reliability. It is conditioned to the observations. In summary, it indicates the ability of the forecast to discriminate between events and non-events given a certain event threshold (WWRP/WGNE, 2013). The area under a ROC curve (**ROCS**) is used as a score. ROCS can take values from 0 to 1, with a value of 0.5 indicating no skill and a value of 1 representing a perfect score. Values lower than 0.5 indicate negative skill. ROC curves measure how good forecasts are in the context of a very simple decision-making model, and are thus better suited to measure how good forecasts are from the perspective of the user than many other commonly used measures (Tveito et al., 2008).

The **BS** [0-1] measures the mean squared probability error and represents the magnitude of the probability forecast errors, with a perfect score of zero. The **Brier Skill Score** (BSS [-∞ to 1]) measures the improvement of the probabilistic forecast relative to sample climatology and indicates what the relative skill of the probabilistic forecast is over that of the climatology, in terms of predicting whether or not an event occurred (WWRP/WGNE, 2013).

The **rank histogram** is used to evaluate whether the forecast ensembles are from the same underlying population as the observations, which implies that the observed would have the same probability of occurrence as any of the ensemble members. This would result in a uniform distribution in the histogram that plots the frequency of the rank of the observation in the ensemble, while deviations from the uniform distribution reveal deficiencies in ensemble calibration, or reliability (Wilks, 2011).

6.2.3.2 Skill assessment

Forecasted streamflow is transformed to the hydrological drought indicator SRI and forecasts of drought are analysed by considering drought conditions to occur for SRI ≤ -0.5 (mild to moderate drought). The value of -0.5 was chosen as it corresponds to the 30th percentile in runoff and it is therefore a good compromise between not capturing all negative anomalies and having a sufficient amount of samples for the analysis. The forecasting system is thus evaluated on the skill of predicting SRI falling below the -0.5 threshold.

However, as we also want to analyse the ability of the system to forecast distributed variables (for agricultural droughts) and water levels in the reservoirs (for irrigation curtailments), we also evaluated the skill of the forecast system in predicting these variables.

6.2.3.3 Estimating uncertainty in the skill scores

Given the small sample size resulting from applying the verification over the 30-year hindcast period, a bootstrap approach is used to estimate the confidence intervals around the ROCS. The idea behind the bootstrap is to treat a finite sample at hand as similarly as possible to the unknown distribution from which it was drawn, which in practice leads to resampling with replacement (Wilks, 2011). The uncertainty of the ROCS is estimated by applying a bootstrap resampling with replacement procedure.

For the FS_S4 and FS_ESP forecasts, we randomly replace (allowing repetition) the original forecast and verification pair to produce a new sample of the same size as our original sample. We then calculate the ROCS from the new sample. We repeat this procedure to create 1000 new samples from which we generate an empirical distribution of the ROCS. The 90% confidence interval is estimated from the 5th and 95th percentiles of this empirical distribution.

For the FS_ESPcond the bootstrap procedure follows the same theory but is computed slightly differently. In this case the bootstrap is achieved by recreating the ensemble forecasts for the hindcast period 1000 times based on the computed probability vector and computing the skill score from each created ensemble.

A limitation of this bootstrap procedure is that statistics computed from discrete bootstrap samples may differ from the ones based on continuous data, and this might lead to overestimation of the confidence. However, this method is widely used in the literature (Dutra et al., 2014; Friederichs and Thorarinsdottir, 2012; Wilks, 2011) to estimate confidence intervals as it does not require assumptions on the distribution.

6.2.4 Assessing spatial hydrological indicators

ROCS and BSS are computed for the spatially distributed indicator root stress (RS) to assess the skill of the forecast in predicting agricultural drought indicators. The RS is an indicator of the available (or the lack of) soil moisture in the root zone, which can be calculated for each grid cell. The RS varies from 0 to 1, where 0 indicates that the soil water availability in the root zone is at field capacity and 1 indicates that the soil water availability in the root zone is at wilting point and the plant is under maximum water stress. For each grid cell, a drought is defined to occur when the root stress is higher than the 70th percentile of the observed values for that month. An advantage of defining the threshold as a percentile of the observed sample as proposed by Roulin (2007) is that it assures a sufficiently large enough number of events to verify and also allows for comparison of verification statistics at different locations (Renner et al., 2009).

In addition to indicators such as RS, it is interesting to evaluate the skill of the model in predicting indicators that are meaningful to the end users in the basin. Irrigation is the major water user in the Limpopo Basin. The amount of water made available to the irrigation sector may, however, be restricted depending on the water level in the reservoirs in the basin as a percentage of their full capacity (DWA, 2013). The forecasted anomaly of the water level in the reservoir is a decision variable that can give an indication to the water managers of the percentage of irrigation demand that can be covered during the season.

An analysis of the historical time series of water level for the Tzaneen reservoir together with the curtailment rules of the reservoir (DWA, 2013) indicate that a 20% curtailment to the irrigation sector is applied when water levels in the reservoir fall below the 50th percentile in the water levels (in percentage of the capacity of the reservoir). Similarly, a 65% curtailment to the irrigation sector is applied when water levels in the reservoir fall below the 37.5th percentile and a 90% curtailment in the irrigation sector along with a 30% curtailment in the urban sector when the water levels are below the 12th percentile. ROCS and BSS are then computed to assess the skill of the forecast in predicting the water levels in the reservoirs to be lower than these threshold percentages of the full capacity. Although the actual operation of the reservoirs is quite a bit more complex, this can be interpreted as an assessment of the skill of the forecast in predicting curtailments to the irrigation sector.

6.3 Results

The following section outlines the results when applying the different types of forcing to the hydrological model over the 30-year hindcast period from 1981 to 2010. The analysis is carried out for different verification periods and lead times as the forecast quality may vary significantly with temporal scales and lead times. While the rainy season in the Limpopo River basins spans from October to March, the main rains typically take place from November to February. The main runoff season and the high runoff season, however, lag behind the rainy season by 1 or 2 months, occurring in general from December to May and from January to April respectively (see Figure 6-3).

6.3.1 Skill of seasonal streamflow prediction

This section presents the skill expressed in the selected skill scores of the seasonal streamflow prediction for the three forecast systems described (FS_S4, FS_ESP, and FS_ESPcond) for Station 24 (Chókwe), Station 1, Station 18 and Station 20 in the Limpopo River basin (see Figure 6-2 for the station locations). Station 24 is the one with the largest drainage area in the basin with available discharge data. Four stations (highlighted in Figure 6-2) with diverse drainage areas were selected to assess the influence of the spatial scale and forecast location on the quality of the forecasts. Table 6-1 presents the main characteristics of these stations, such as drainage area, mean annual runoff and observed runoff coefficient (RC = runoff/precipitation). In these stations the performance of the hydrological model is found to be satisfactory based on the evaluation measures and ranges proposed by Moriasi et al. (2007), which comprise the Nash-Sutcliffe efficiency (NSE), and the ratio of the root mean square error to the standard deviation of the measured data (RSR). The coefficient of determination (R^2) is also included. These results are presented in Section 4.4.1 and are summarised in Table 6-1.

Table 6-1 Model evaluation measures for runoff for selected stations, ordered by basin size.

Station number	Subbasin area (km²)	Mean annual observed runoff (m³ s⁻¹)	RCobs (%)	R^2	NSE	RSR
24	342,000	96.9	1.7	0.92	0.90	0.32
1	201,001	39.5	1.2	0.69	0.57	0.65
18	98,240	12.2	0.7	0.68	0.62	0.62
20	12,286	14.8	5.3	0.70	0.65	0.59

Figure 6-4 (upper plots) presents the ROC diagram for the 6-month SRI-6 ≤ -0.5. For calculating the SRI-6 the verification period is from December to May and the SRI-6 value is recorded at the end of the period in May. The figure shows three of the four stations considered, for a lead time of 5 months (the forecast is issued in December). December is the only start time of the forecast that captures the whole 6-month main runoff season (from December to May) in the seasonal forecast. The ROC diagram for Station 18 is not presented given that it has a similar behaviour to Station 1. The ROC curves are presented for each forecasting system, and the ROC of FS_ESPcond is represented by the ensemble that results in the median ROCS. Results from the FS_ESPcond show for all stations a narrower 90% confidence interval when compared with the other two forecasting systems considered (see middle and lower plots in Figure 6-4), thus suggesting that FS_ESPcond is more robust. Histograms of ROCS for FS_ESP are not shown as these are similar to those of FS_S4.

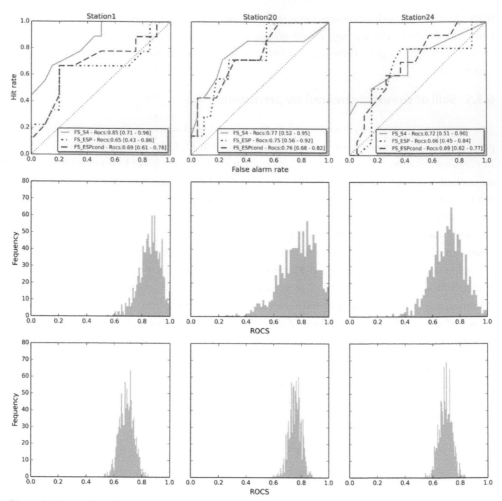

Figure 6-4 Upper plots: Relative operating characteristic (ROC) diagram representing false alarm rate versus hit rate for the 6-month SRI (DJFMAM) ≤ -0.5 given by FS_S4, FS_ESP and FS_ESPcond for three stations (1, 20, and 24). The ROCS for each forecasting system together with the 90% confidence interval (5-95th percentiles) resulting from the bootstrap are indicated in the legend. Middle and bottom plots: histogram of the bootstrapped ROCS for FS_S4 (middle) and FS_ESPcond (lower) respectively for the same three stations.

The ROCS of the FS_S4 in predicting SRI-6 ≤ -0.5 are generally quite high (around 0.8), but some lower values such as 0.72 (this for the station with largest contributing area) are observed (Figure 6-4). The lower skills for the station with the largest contributing area for FS_S4 might be attributed to the shift from an arid to a more tropical climate, which means that the persistence of initial conditions would be lower. Also, given that this is mostly the case for the FS_S4 and less so for the FS_ESP and FS_ESPcond, we can speculate the ECMWF S4 seasonal forecast might have a better skill for the northern (more arid) part of the basin (area corresponding to the sub-basin draining to Station 1), than for the southern part of the basin. FS_ESP generally shows the lowest skills, with the skills of FS_ESPcond in between FS_ESP and FS_S4. The verification was also done for forecasts issued for the 4-month period JFMA (high-runoff season) with forecasts issued from November to January, respectively. Figure 6-5 presents the ROCS for the 4-monthly SRI (SRI-4 in April) ≤ -0.5 for three different lead times (3–5 months) and two stations.

For the high runoff season SRI-4, similar results to those of SRI-6 are observed. In almost every case FS_S4 shows higher skill than FS_ESP and FS_ESPcond. The skill of the forecasts tend to decrease with lead time, especially for FS_ESP and FS_ESPcond, which do not show any skill at the 5-month lead time. In contrast, the skill of FS_S4 for the 5-month lead time is still good. The skill score verification for SRI-1 for the same 4 months period January-April (not shown) shows once more that the FS_S4 is more skilful than the other two forecasting systems. The smaller subbasins (18 and 20) present lower skill for SRI-4 for the three forecasting systems, and while subbasin 18 still presents some skill for all the three FS, subbasin 20 only does so for the FS_S4 for all lead times. In general, for all locations, the skill of the FS_S4 decreases slightly with lead time, while the skill of both FS_ESP and FS_ESPcond decreases more rapidly with lead time. A curious fact is that for a few stations (e.g. 1, 18), for both SRI-4 and SRI-1, the FS_S4 shows a higher skill for a lead time of 4 months than for a lead time of 3 months. For the SRI-4, this means that the forecast is more skilful in predicting the April SRI-4 when issued in December than when issued in January. However, the differences are not statistically significant and this can be due to sampling errors.

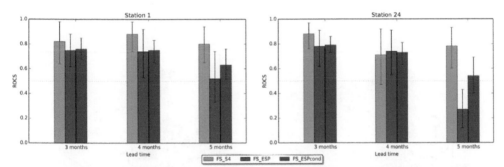

Figure 6-5 ROCS for the SRI-4 (JFMA) ≤ -0.5 given by FS_S4, FS_ESP, and FS_ESPcond for different lead times, for two of the stations (1, 24). The error bars represent the 5-95th percentiles of the bootstrapped ROCS values.

Rank histograms for every station and lead time together with the results of the Kolmogorov–Smirnov test show that for the three forecasting systems, uniformity of the distribution cannot be rejected, indicating the forecasts are reliable. Figure 6-6 presents the rank histograms of SRI-6 for Station 1 for the three forecasting systems as an example.

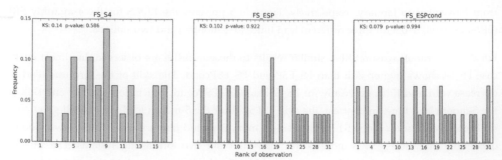

Figure 6-6 Rank histograms of SRI-6 for Station 1 for the three forecasting systems (FS_S4, FS_ESP, and FS_ESPcond). The results of the Kolmogorov–Smirnov test for uniformity are presented in each plot.

6.3.2 Skill of spatial hydrological indicators

Figure 6-7 shows the ROCS and the BSS of the FS_S4 in predicting agricultural drought conditions, i.e. in predicting aggregated RS during the 6-monthly period DJFMAM to be higher than the 70th percentile. Yuan et al. (2013) show that the annual cycle of soil moisture in southern Africa (simulated by the VIC model) lags behind the precipitation. Figure 6-7 shows that the skill of the FS_S4 forecast in predicting agricultural droughts is higher than climatology (ROCS > 0.5, BSS > 0) throughout the entire basin.

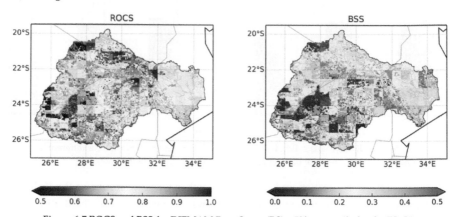

Figure 6-7 ROCS and BSS for DJFMAM Root Stress (RS) > 70th percentile for the FS_S4.

To assess the skill of the seasonal forecast in predicting a specific decision variable in the Limpopo River basin, we calculate the skill of the forecast in predicting water level thresholds in the reservoir that would result in curtailment to the irrigation sector. The availability of water is represented in each cell by the water level. In the cells corresponding to reservoirs, the water level is a surrogate for the storage and is described as a percentage of the full storage capacity of the reservoir. Figure 6-8 presents the ROCS and the BSS of the FS_S4 in predicting water levels during the 6-monthly period DJFMAM to be lower than the 50th and 37.5th percentiles, based on the analysis described in section 6.2.4. The figure shows that the skill of the FS_S4 forecast in predicting low water levels is higher than climatology (ROCS > 0.5, BSS > 0) throughout the basin. The spatial distribution across the basin does show the skill to be higher in the northern basin than in the southern basin, which may contribute to the lower skill found at Station 24 close to the basin outlet than at Station 1 in the upper (northern) basin.

The skill scores in cells that contain the reservoirs are represented by a circle to enhance visibility. It is clear from Figure 6-8 that the skill of the forecast in predicting low water levels is higher in the reservoirs than in nearby streams. This can of course be expected due to the higher memory introduced by the reservoir's storage capacity with respect to the streams. Figure 6-9 presents the forecast probability of water levels to be lower than the 50th and 37.5th percentiles during the Dec 1991- May 1992 season as an example. This was the driest season in the last 30 years. The forecast is issued in December 1991.

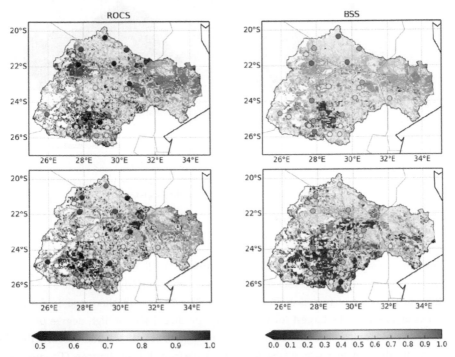

Figure 6-8 ROCS and BSS for: Water Level (WL) < 50th percentile (upper plots), and WL < 37.5th percentile (lower plots) for the FS_S4.

The forecast probability of water levels in the reservoirs being lower than the 50th and 37.5th percentiles can be interpreted as the forecast probability of a curtailment of 20% and 65%, respectively in the irrigation sector during the season. For several reservoirs in the basin the FS_S4 forecast issued in December 1991 predicted a high probability of curtailment to the irrigation sector during the Dec1991-May1992 season. Records confirm the lower than normal water levels during this season, with the irrigation quota indeed being curtailed (DWA, 2013).

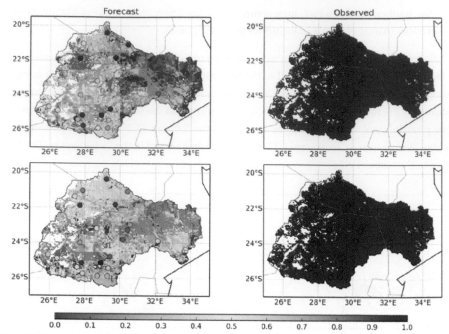

Figure 6-9 Forecast probability of water level (WL) < 50th percentile (upper plots), and WL < 37.5th percentile (lower plots) for the FS_S4 during the season Dec 1991- May 1992 issued in Dec 1991 (left panels), and what actually occurred: 1=yes, 0=no (right panels).

6.3.3 Analysis of a specific event

Yuan et al. (2013) note that "The major source of seasonal forecast predictability comes from the ocean, and the strongest signal is the El Niño-Southern Oscillation (ENSO)". Given that the ECMWF S4 is influenced by the ENSO signal, it is interesting to analyse how the FS_S4 predicts streamflow in the onset of two clear El Niño years. The 1997/98 El Niño year is described in Thomson et al. (2003) as the largest for this century, predicted with a high degree of certainty. Although many of the climate anomalies typical of an El Niño event took place around the globe, the devastating drought that was feared for southern Africa did not happen (Thomson et al., 2003). For this analysis another year was selected that had a less strong ONI but that did result in a severe drought (1982/83). Figure 6-10 presents the ensemble seasonal streamflow prediction from FS_S4 for both the 1997/98 and the 1982/83 seasons issued in October and updated in December for Station 24. The plots also show the climatology of the streamflow and the 30th percentile, i.e. the value below which 30 percent of the observations are found. The reference streamflow for that season and the forecast ensemble mean are also shown.

Figure 6-10 shows that in October the predictions from the forecasting system FS_S4 for El Niño seasons of 1982/83 and 1997/98 were relatively similar (see Figure 6-10 upper panels) even though the 1997 JAS ONI was notably higher than the 1982 JAS ONI. The updated forecast in December, however, shows a different situation: While the forecast for the 1982/83 season point towards very dry conditions, the forecast of the 1997/98 season indicates near-normal conditions. Yet, the 1997 SON ONI is markedly higher than the 1982 SON ONI. Thus, in spite of the strong ONI conditions, the S4 system correctly forecasted the no-drought condition in the 1997/98 season. This indicates that even though the S4 forecasting system is influenced by the sea surface temperatures over the Niño-3.4 region, the precipitation and temperature forecasts over the Limpopo region are not only constrained by the sea surface temperature evolution, but results from the atmospheric circulation response to different climate forcing.

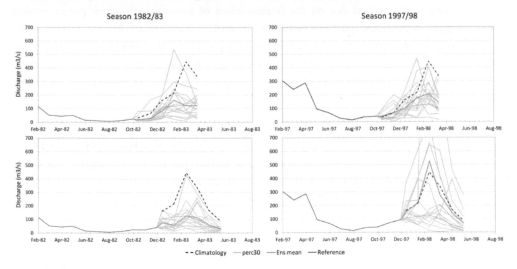

Figure 6-10 Seasonal forecast FS_S4 for 2 seasons issued in October (upper panel) and December (lower panel).

6.4 Discussion

The performance of the three hydrological forecasting systems constructed with the same hydrological model and different meteorological ensemble forecasts are evaluated by means of widely used probabilistic verification skill scores, including the ROC diagram and the rank histogram. Among the forecasting systems considered in this study, FS_ESP is considered the most traditional. Such traditional approaches for hydrological forecasts rely on historical observations of the meteorological conditions, without considering meteorological forecasts. In ensemble probabilistic forecasting, the ESP approach, implicitly accounting for hydrologic persistence and historical variability of climate, is normally used (Brown et al., 2010). FS_S4 is a more complex forecasting system as it requires as forcing the outputs of a seasonal meteorological forecast system, which are complex numerical models and resource intensive. FS_ESPcond, a modification of the ESP approach, conditions its ensemble on past years that had similar climate conditions to the year in which the forecast is made (Brown et al., 2010). Given that the Limpopo region is known to be affected by ENSO and droughts tend to occur during El Niño years, the forecast ensemble was constructed by assigning weights to the different ensemble traces based on the El Niño index.

The skill evaluation of the seasonal forecasts is limited by the use of model data as verification, i.e. we verify our forecasts against the baseline simulation, which was also used to provide the initial conditions to the forecasts. This is the same approach as taken in Yossef et al. (2013), Winsemius et al. (2014), Shukla et al. (2013) and Renner et al. (2009), and while it allows for the detailed (spatial) evaluation of the skill of the forecasts, it can potentially hide limitations of the modelling system. Therefore, these skill results should be interpreted as the upper limit of real predictability of the current system. Biases in streamflow prediction can often be removed through statistical post-processing. Based on the historical performance of the forecasting system, operational streamflow forecasts are statistically corrected in real-time (Verkade et al., 2013). Several studies were carried out on the improvement of forecast skill when pre-processing meteorological forcings and post-processing of streamflow (Kang et al., 2010; Zalachori et al., 2012; Yuan and Wood, 2012). They found that both pre-processing and post-processing improved the forecast skills, and have a limited contribution to reduction of the forecast uncertainties (Verkade et al., 2013). However, Verkade et al. (2013) found that the improvements to streamflow forecasts accrued by pre-processing were modest.

Results of the seasonal streamflow prediction show that for every lead time FS_S4 is skilful in predicting SRI-6, SRI-4, and SRI-1 during the summer rainy season, while for the other FS the skill is lower and decreases more rapidly with lead time. This means that the complex S4 seasonal forecasting system adds value to the hydrological predictions compared to the climatology-based forecasting systems, as well as the ENSO-mode-conditioned climatology forecast systems. This was also observed during a specific event where expected anomalies due to El Niño did not materialise, but FS_S4 detected this. The skill decreases when going from SRI-6 to SRI-4 and SRI-1. This is as expected given the higher variability of the predictand for shorter aggregation periods. The skill from FS_ESP is lower than that of FS_S4 in almost every case, while the skill of FS_ESPcond is in general between the other two. For SRI-4, FS_ESP and to a lesser extent FS_ESPcond do not show any skill for 5-month lead time at any of the stations considered.

As expected, the skill of all forecasts tends to decrease with lead time. This is, however, especially the case for FS_ESP and FS_ESPcond where the decrease in skill with lead time is larger than for FS_S4. For the smaller aggregation periods (SRI-4 and SRI-1) FS_ESP deteriorates to climatology already at a lead time of 3 months for stations 18 and 20, the upstream basins of which are smaller in size. In the larger basins FS_ESP shows predictability up to a 4-month lead time, probably due to the spatial aggregation taking place over larger basins smoothing out uncertainties in space. This indicates that the memory in the hydrology (storage in groundwater, reservoirs, channels and wetlands) contributes to the predictability with a lead time of up to 2–4 months. For longer lead times, the meteorological forcing dominates the predictability of the system. The critical lead time after which the importance of the meteorological forecast exceeds that of the initial conditions depends on the location and size of the basin and should be analysed for each sub-basin of interest. Rank histograms for every station and lead time indicate that the three forecast systems are reliable given that uniformity of the distribution cannot be rejected.

6.4.1 What does the analysis mean to end users?

The high predictability of FS_S4 for all lead times and aggregation periods of SRI is encouraging given that such a system, if made operational, may provide end users with sufficient time to decide upon measures to take in anticipation. For example, they might decide to change the cropping date or the cropping area if they expect not to have enough water to fulfil the crop requirements. Therefore, there is added value to using a seasonal meteorological forecast (ECMWF S4) to force the hydrological forecasting system when compared to the conventional ESP. The higher skill of the FS_S4 and FS_ESPcond compared to that of the FS_ESP for every lead time is in line with the study of Shukla et al. (2013), who show that for the region of the Limpopo River basin the meteorological forecast dominates the hydrological predictability for the wet season for almost every lead time considered. Only for the 1-month lead time forecasts issued in October did they find a higher influence of the hydrological initial conditions to some extent. Moreover, Yossef et al. (2013) indicate that for semi-arid regions the initial conditions do not contribute much to the skill given the high sensitivity of the runoff coefficient to rainfall variability.

The FS_S4 was also evaluated regarding its ability to predict agricultural droughts and curtailments in irrigation (water levels lower than the 50th and 37.5th percentiles). Maps of spatially distributed ROCS and BSS (Figures 6-7, 6-8) show that the skill of the FS_S4 forecast in predicting these conditions is higher than climatology (ROCS > 0.5, BSS > 0) throughout the basin. Indicating the probability of curtailment to the irrigation sector during the following season is an example of providing a forecast in an understandable format that is useful to the end users. If they are informed that there is a high probability of a high curtailment to the water available for irrigating their crops during the following season, users would have a clear idea of what is the best practice for that situation. Further improvements in forecasting skill could be achieved through better meteorological predictions or better estimation of initial conditions (Yossef et al., 2013). Whether the forecasts indeed have value will depend on the costs of decisions made in response to the forecast, losses in case of a wrong decision and the gain in case of a good decision. This should be further analysed in a continuation of this study.

As a next step, it is recommended that the forecast skill of the FS_S4 and FS_ESPcond be assessed in an actual forecasting mode for a following summer season. The seasonal meteorological forecast from S4 can be obtained in real time for research purposes. To test a pre-operational system, the forecasting system ought to be statistically post-processed in order to remove biases in streamflow predictions. Moreover, the initial conditions for the forecasts could be better estimated through data assimilation of water levels in reservoirs and streams. This data could be obtained from the water managers of the basin. Despite the limitations of FS_S4 (access to real-time atmospheric-ocean seasonal forecasts for non-ECMWF member-states, and their quality) and FS_ESPcond (depending on the calibration and decreased skill at long lead times), both systems show potential for seasonal hydrological drought forecasting in the Limpopo River basin to provide operational guidance to users.

6.5 Conclusions

We evaluate the performance of three forecasting systems (FS_S4, FS_ESP, and FS_ESPcond) in the Limpopo River basin. These systems make use of the same hydrological model and are forced with three different meteorological ensemble forecasts (two of which are based on resampled climatological records, FS_ESP and FS_ESPcond, and one based on seasonal meteorological forecasts, FS_S4). Results of the seasonal streamflow prediction show that the three forecasting systems show moderate skill in predicting SRI-6 (DJFMAM) ≤ -0.5. Moreover, the three forecasting systems are unbiased as suggested by the rank histograms.

For every lead time and aggregation period considered, FS_S4 is found to be skilful in predicting hydrological droughts represented by SRI ≤ -0.5 during the summer rainy season. The skill decreses when going from SRI-6 to SRI-4 and SRI-1, as well as with increasing lead time. The skill of FS_ESP is lower than that of FS_S4 in almost every case and deteriorates rapidly with lead time, showing no skill after a lead time of 4-5 months for SRI-4 and SRI-1. This indicates that the memory in the hydrology contributes to the predictability up to 2-4 months but for longer lead times the predictability of the system is dominated by the meteorological forcing. FS_ESPcond shows in general lower skills than FS_S4 but it becomes comparable and can even outperform the latter for smaller lead times if the parameters for selection and weighting of ensemble members are carefully calibrated for each basin. Moreover, the skill of FS_ESPcond is more robust than that of the other forecasting systems as suggested by the narrower confidence intervals of ROCS. As with FS_ESP, the skill of FS_ESPcond also decreases faster than that of FS_S4 with lead time.

The high predictability of drought of FS_S4 for all lead times and aggregation periods of SRI and for the spatial drought indicators is encouraging given that such a system, if made operational, may provide end users with sufficient time to decide upon measures to take in anticipation. Moreover, FS_ESPcond shows promising results. This forecasting system only requires the ONI index previous to the forecast to weight the ensemble traces to include in the forecast. This system is relatively simple and presents the advantage that it can be coupled with the forecast of the ONI index that is available with a long lead time. Naturally, in this situation the uncertainties of both forecasts need to be considered.

7

CONCLUSIONS

7.1 General

Among all natural hazards, droughts have resulted in the highest number of deaths worldwide causing more casualties than all the other hazards combined, as well as damage to the environment and ecosystems. Moreover, social impacts include conflicts and wars, migration or relocation, and increase of water-borne diseases (West, 2014). Africa has suffered from acute droughts in the past, which contributed to food insecure conditions in several African countries and demanded crisis management. The United Nations Development Programme estimated that around 220 million people were found to be exposed annually to drought and African states were found to be the most vulnerable to drought (UNDP, 2004).

Water shortage at local and regional scale is foreseen to intensify as a result of climate variability, climate change, and human pressure. For these reasons, drought management should move towards a more proactive risk management approach in order to improve resilience and preparedness to drought. Improving resilience and preparedness is crucial to reduce drought vulnerability and risk to the societies. This is currently being promoted by the World Meteorological Organization and the United Nations Convention to Combat Desertification, which are encouraging the development of drought policies and the evolution of the drought management approach from crisis management to proactive risk management (UNCCD et al., 2012; Pozzi et al., 2013). A proactive drought risk managements requires and effective drought early warning system in order to provide users with sufficient lead time to put mitigations strategies into practice.

This research was carried out in the framework of DEWFORA project and focuses on the development of a modelling framework for hydrological drought forecasting in sub-Saharan Africa as a step towards an effective early warning system.

7.2 Main contributions

This research has high scientific and societal significance. Scientifically, it contributes to a better understanding of the tools needed for hydrological drought forecasting in the African continent for different time horizons and spatial scales. This research focuses on the forecasting of an effective early warning system. On the societal aspect, results of this research contributed to the development of a course on drought forecasting which aims to transfer the knowledge developed to practitioners and develop capacity in Africa and other regions. A more effective drought forecasting and warning system will hopefully contribute to important aspects in the region such as food security, hazard management, and risk reduction.

A key contribution of this PhD research to improving drought forecasting capabilities in Africa is the development of a seasonal probabilistic hydrological drought forecasting system for the Limpopo River basin. The forecasting system consists of a distributed hydrological model forced with available state-of-the-art meteorological forecasts. Even though the system was developed for the Limpopo Basin, the method can very well be adapted to other basins, e.g. the adjacent Incomati basin, but also on a continental level. For example, the continental hydrological model developed during this research is currently used in the African Drought Observatory (ADO) hosted by JRC (http://edo.jrc.ec.europa.eu/ado/ado.html). Building the probabilistic forecasting system for the Limpopo Basin required several clear steps which were presented in the different chapters. The main conclusions are here summarised.

The first step was to select an appropriate continental hydrological model for the forecasting system. Several widely used hydrological models were reviewed to assess their suitability for drought forecasting in Africa. A framework for selecting models for drought forecasting was presented and used to select a subset of models (from a larger set) that are considered suitable for drought forecasting, in some cases assuming some possible adaptations. The suitability of the models was assessed through applying a set of criteria such as the representation of most relevant physical processes, applicability of the model for operational use for drought early warning, and the capability of the model to be downscaled to a smaller scale. Among sixteen hydrological and land surface models selected for this review; PCR-GLOBWB, GWAVA, HTESSEL, LISFLOOD and SWAT showed higher potential and suitability for hydrological drought forecasting in Africa based on the criteria used in this evaluation. Among these models, the PCR-GLOBWB was chosen for further application in this research.

The PCR-GLOBWB model was set up for the entire African continent. Given that the main purpose of our model setup was to simulate droughts, or more specifically drought indicators, we centred our attention on the representation of actual evaporation, which is an important component for drought assessment. The possibilities to validate a continental evaporation product for Africa are unfortunately limited due to the inexistence of a continental-scale evaporation product based on ground measured data. However, in recent years an increasing amount of studies have focused on global evaporation estimates, and a number of evaporation estimates or products have been developed. The main contribution of Chapter 3 is an evaporation analysis focused on the African continent which serves as an indirect validation of methods or tools used in operational water resources assessments. Our analysis discriminates areas where there is good consistency between different evaporation products and areas where they diverge. In some regions such as the southern Africa, the agreement between the products was found to be very good, which means that use of any product may be equally good for developing drought indicators. In other regions, such as in the humid Sahel or in the Mediterranean area the choice of evaporation product needs to be further studied as there is a large difference between the products. We also presented a range of variance in actual evaporation that can be expected in a given region, which may be useful in, for example, water resources management when estimating the water balance. We derived an Actual Evaporation Multiproduct at a 0.5° resolution that integrates satellite based products, evaporation results from land-surface models and from hydrological models forced with different precipitation and potential evaporation data sets, and may serve as a reference (benchmark) data set for Africa.

After testing the hydrological model at the continental scale, we narrowed our study area to the Limpopo River basin, one of the most water-stressed basins in southern Africa. We applied a higher resolution version of the same hydrological model to the Limpopo Basin and analysed the performance of the model in simulating space-time variability of historical droughts in the basin, expressed through selected drought indicators. Most of the indicators considered were able to represent the most severe droughts in the basin and to identify the spatial variability of the droughts. We found that even though meteorological indicators with different aggregation periods may be used to characterise different types of droughts reasonably well, there is added value in computing indicators based on the hydrological model for the identification of droughts and their severity. For example, even though the Standardized Precipitation Index (SPI) can give a reasonable indication of hydrological drought conditions, computing the Standardized Runoff Index (SRI) can be more effective for the identification of hydrological drought. On the other hand, in the absence of actual evaporation and soil moisture data, the meteorological indicator Standardized Precipitation Evaporation Index aggregated for a 3-months period (SPEI-3), which considers both precipitation and potential evaporation and is reasonably easy to compute, may be used as an indicator of agricultural droughts. A combination of different indicators, such as SPEI-3, SRI-6, and SPI-12, can be an effective way to characterise from agricultural to long-term hydrological droughts in the Limpopo river basin.

We then discussed spatial scale in hydrological modelling based on the two resolution versions of the hydrological model: a "low" resolution (0.5° × 0.5°) hydrological model, which covers the whole of Africa, and a "high" resolution (0.05°× 0.05°) model for the Limpopo River basin. Here we investigated if the processes that are normally smoothed out in the global models can be recaptured by downscaling the resulting hydrological variables from a continental hydrological model to finer spatial scales. Using the results of the two resolution models, we investigated the effect of spatial resolutions on three distributed fluxes or storages: actual evaporation, soil moisture, and total runoff. We used the coefficient of variation (CV) to assess the variability of the high resolution variables within each low resolution pixel. Dry seasons and dry years were identified with higher CV than wet seasons and years. In a next step, we applied two techniques to downscale hydrological variables; bias correction statistical downscaling, and downscaling by using topographic and soil attributes as explanatory variables. The results from the two downscaling methods are similar. The downscaling results for evaporation and soil moisture are reasonably good, particularly for the wet season, but the results for the runoff are not as good. The analysis of the variability of the fluxes on high resolution grid cells under different land features and soils indicated that there is good potential of downscaling the low resolution hydrological model results to high resolution based only on the terrain and soil characteristics.

Lastly, we evaluated the performance of three forecasting systems in the Limpopo River basin. These three forecasting systems make use of the high resolution hydrological model, but are forced with three different ensemble meteorological forecasts. Two of these are based on resampled climatological records, and one uses a seasonal meteorological forecast product (FS_S4). For every lead time and aggregation period considered, FS_S4 was found to be skilful in predicting hydrological droughts during the summer rainy season. Results show that the persistence of the initial hydrological conditions contributes to the predictability for up to 2 to 4 months, while for longer lead times the predictability of the system is dominated by the meteorological forcing. The predictability of drought by FS_S4 for all lead times and aggregation periods of SRI, as well as for spatial drought indicators is encouraging. Results of the two forecasting systems that are based on resampled climatological records demonstrate that when using the resampled historical records the lead time at which the forecasts are skilful is limited to the persistence of the initial states. However, if the historical samples are weighted conditional on the ENSO signal as expressed by the ONI index at the start of the forecast the lead time improves significantly. This forecasting system is quite a bit less complex than using the FS_S4 forcing.

If made operational, these hydrological forecasting systems may provide end users with useful information at sufficient lead time. This will help them to decide upon measures to take in anticipation of drought events, thus, reducing the impacts of those events.

7.3 Recommendations for future research

Although the research shows the significant potential of model based hydrological drought forecasting in Africa, the system developed in this study contains several limitations that should be further addressed to improve reliability of the forecasts.

In this research the PCR-GLOBWB (1.1) hydrological model was found to be suitable for drought forecasting in Africa during the initial review and evaluation, and this was further confirmed in the following parts of the research where the model was evaluated and used for the seasonal forecast of hydrological droughts. However, the PCR-GLOBWB model is under a continuous process of improvement by the development team and several advances have been made in the last few years. PCR-GLOBWB 2.0 (currently under development) will operate at 5 minutes globally and will fully integrate water demand calculations, abstractions and return flows. Moreover, it will have a coupled groundwater flow model. It is recommended to evaluate whether the improvements made to the representation of hydrological processes will result in better forecast skill, and at which scale or physiographic setting.

This is particularly relevant for groundwater flow resulting from the model, which may not be fully adequately represented in the Limpopo Basin using the current version of the model. Groundwater is an important component for drought assessment, but was not studied in depth due to constraints such as limited information on the hydrogeology and groundwater tables in the Limpopo River basin. Moreover, several studies suggest that groundwater in the Limpopo is being depleted (Wada et al., 2010; FAO, 2004). In Botswana, "... groundwater is considered a non-renewable resource because of the very low recharge rates, and will only be used in cases of emergency to augment existing surface water supplies in Gaborone" (FAO, 2004). Across Africa, the aquifers are heterogeneous and discontinuous. Hence, the availability and distribution of groundwater resources over the continent is complex and not well understood. In semi-arid areas, the storage can be considerable but it is mostly non-renewable as recharge can be very low. Vast amounts of groundwater are stored in non-renewable reserves across the continent in sedimentary formations. In addition, the distribution of renewable groundwater over the continent is highly skewed, with more than half of its renewable groundwater contained in just four countries (UNECA-ACPC, 2011). It is recommended that further research on drought assessment in the Limpopo, as well as over whole of Africa, emphasises on the role of groundwater in monitoring and forecasting hydrological droughts.

Additionally, the comparison of continental evaporation and the Actual Evaporation Multiproduct (EM) should be extended to validate the products in different African regions with ground data, despite its limited availability. Other available products should also be added to the comparison and to the derivation of the EM to provide more information on the variance between the products and a more consistent EM estimate. It is also recommended to compare the computed EM and the variability of the products with global benchmarks such as that recently developed by Mueller et al. (2013). Similarly, at the basin-wide scale, long-term estimates of evaporation could be obtained from the water balance with an uncertainty estimate (Dingman, 1994).

The evaluation of the skill of our seasonal forecasts was limited by the use of model data as verification, i.e. we verify our forecasts against the baseline simulation, which was also used to provide the initial conditions to the forecasts. This approach allows for the detailed spatial evaluation of the skill of the forecasts, but can potentially hide limitations of the modeling system as it factors out uncertainties in the hydrological model. While the same approach is taken in Yossef et al. (2013), Winsemius et al. (2014), Shukla et al. (2013), Verkade et al. (2013), and Renner et al. (2009), the results of the skill assessment should be interpreted as the upper limit of the real predictability of the current system. Biases in streamflow prediction can often be removed through statistical post-processing. Based on the historical performance of the forecasting system, operational streamflow forecasts can be statistically corrected in real-time (Verkade et al., 2013).

An additional constraint on the skill evaluation is that when a single hydrological model is used to investigate ensemble flow predictions from meteorological ensembles as forcing, only the uncertainty originating from weather predictions is assessed (Velázquez et al., 2011). It is well known that there are other types of uncertainty to be considered, such as the uncertainty of the model parameters and structure. However, these uncertainties are considered outside the scope of the present study, but do warrant further research. Finally, whether the forecasts indeed have value will depend on the decisions made in response to the forecast, and the ratio of the costs of those decisions to the losses in case of a wrong decision as well as the gain in case of a good decision. This should be further analysed in a continuation of this study.

It was beyond the scope of this PhD research, and beyond the scope of the DEWFORA project, to develop a fully operational drought forecasting system. However, we see that the research developed will contribute to a future operational drought early warning system for Africa. However, it needs to be carried forward by the organisations that have a mandate for providing operational drought forecasts. Nevertheless, the state-of-the-art seasonal meteorological forecasts have been developed during the DEWFORA project and are now available for Africa. Despite the notable advances on drought monitoring and forecasting in Africa, the use of forecasting tools in decision making is still limited, in some cases by the response capability. For instance, the 2010/11 drought in the Horn of Africa was well predicted by ECMWF, but this information was not timely used for better preparedness and mitigation of the drought, which finally caused a heavy toll affecting about 12 million people (Dutra et al., 2013).

On a user level, despite seasonal meteorological forecasts being available in the Limpopo Basin, farmers currently seem to prefer local indigenous and traditional knowledge drought forecasting systems. This is a clear example where education and capacity development can help local users to understand the hydro-meteorological forecasts and thus increase the response capability of the early warning system. Moreover, it is recommended to research on how to integrate the local indigenous knowledge to the developed seasonal forecasting system to increase the forecast skill.

This shows that even if the hydrological drought forecasting system for the Limpopo Basin is sufficiently mature to become operational, there are several challenges that need to be addressed for the resulting seasonal forecasts to be effective for the local community. Dedication of time, training, and established participative measures with policy-makers and water managers are necessary for the system to be adopted. Furthermore, end users should receive the information in an understandable format at the time they need it for the forecast to be useful. The highly technical information that is typically contained in the forecasts should then be translated to a comprehensible form before being disseminated and delivered to decision makers and water users. This will require additional research on the use of such forecasts products, and how those could be tailor made to meet specific user needs. Additionally, end users should be involved in the forecast verification by providing feedback to the forecasters (DEWFORA, 2012b). Lastly, education and capacity development are prerequisite the success and sustainability of any early warning system.

Finally, it is highly recommended that collaboration between different groups that are working on drought forecasting both in Africa and worldwide is established. For example, collaboration should be established between the current research and other efforts such as the Princeton African Flood and Drought Monitor. Recently, progress has been made towards a Global Drought Early Warning Monitoring Framework (GDEWF), and the establishing of a Global Drought Early Warning System (GDEWS) as an interoperable information system (Pozzi et al., 2013). The improvements on regional drought forecasting capabilities in southern Africa developed in this research contribute to the objectives set out in the GDEWF, so it is therefore recommended that this research as well as its continuation is linked to those efforts.

8

REFERENCES

Abebe, N. A., Ogden, F. L., and Pradhan, N. R.: Sensitivity and uncertainty analysis of the conceptual HBV rainfall-runoff model: Implications for parameter estimation, Journal of Hydrology, 389, 301-310, 2010.

Abramowitz, M., and Stegun, I. A.: Handbook of mathematical functions, with formulas, graphs, and mathematical tables, Dover Publications, 1046 pp., 1965.

Allen, R. G., Pereira, L., Raes, D., and Smith, M.: Crop evapotranspiration - Guidelines for computing crop water requirements: FAO Irrigation and drainage paper No. 56, FAO, Rome, 26-40, 1998.

Alley, W. M.: The Palmer Drought Severity Index: Limitations and Assumptions, Journal of Climate and Applied Meteorology, 23, 1100-1109, doi: 10.1175/1520-0450(1984)023<1100:tpdsil>2.0.co;2, 1984.

Alston, M., and Kent, J.: Social impacts of drought, Centre for Rural Social Research, Charles Sturt University, Wagga Wagga, NSW, 2004.

Alton, P., Fisher, R., Los, S., and Williams, M.: Simulations of global evapotranspiration using semiempirical and mechanistic schemes of plant hydrology, Global Biogeochemical Cycles, 23, GB4023, doi: 10.1029/2009GB003540, 2009.

AMS: Meteorological drought - Policy statement, Bulletin of the American Meteorological Society 78, 3, 1997.

AMS: American Meteorological Society, Meteorological Drought (Adopted by AMS Council on 23 December 2003), Bull. Amer. Met. Soc., 85, https://www.ametsoc.org/policy/droughstatementfinal0304.htmll, access: February 2015, 2004.

AMS: Drought - An Information Statement of the American Meteorological Society (Adopted by AMS Council on 19 September 2013), American Meteorological Society, https://www.ametsoc.org/POLICY/2013drought_amsstatement.pdf, 2013.

Armstrong, R. L., Brodzik, M. J., Knowles, K., and Savoie, M.: Global monthly EASE-Grid snow water equivalent climatology, Boulder, CO: National Snow and Ice Data Center, Digital media, 2007.

Balsamo, G., Beljaars, A., Scipal, K., Viterbo, P., van den Hurk, B., Hirschi, M., and Betts, A. K.: A Revised Hydrology for the ECMWF Model: Verification from Field Site to Terrestrial Water Storage and Impact in the Integrated Forecast System, Journal of Hydrometeorology, 10, 623-643, 2009.

Balsamo, G., Boussetta, S., Lopez, P., and Ferranti, L.: Evaluation of ERA-Interim and ERA-Interim-GPCP-rescaled precipitation over the USA, ECMWF ERA Report Series 5, available at: http://old.ecmwf.int/publications/library/ecpublications/_pdf/era/era_report_series/RS_5.pdf, last access: August 2014, 1-25, 2010.

Balsamo, G., Boussetta, S., Dutra, E., Beljaars, A., Viterbo, P., and Hurk, B.: Evolution of land-surface processes in the IFS, ECMWF Newsletter 127, Spring 2011, 2011a.

Balsamo, G., Pappenberger, F., Dutra, E., Viterbo, P., and van den Hurk, B.: A revised land hydrology in the ECMWF model: a step towards daily water flux prediction in a fully closed water cycle, Hydrological Processes, 25, 1046-1054, 2011b.

Balsamo, G., Albergel, C., Beljaars, A., Boussetta, S., Brun, E., Cloke, H. L., Dee, D., Dutra, E., Pappenberger, F., Munoz Sabater, J., Stockdale, T., and Vitart, F.: ERA-Interim/Land: A global land-surface reanalysis based on ERA-Interim meteorological forcing, ECMWF ERA Report Series, 13, 1-25, 2012.

Balsamo, G., Albergel, C., Beljaars, A., Boussetta, S., Brun, E., Cloke, H., Dee, D., Dutra, E., Muñoz-Sabater, J., Pappenberger, F., de Rosnay, P., Stockdale, T., and Vitart, F.: ERA-Interim/Land: a global land surface reanalysis data set, Hydrol. Earth Syst. Sci., 19, 389-407, doi:10.5194/hess-19-389-2015, 2015.

Barbosa, P., Naumann, G., Valentini, L., Vogt, J., Dutra, E., Magni, D., and De Jager, A.: A Pan-African map viewer for drought monitoring and forecasting, 30 October to 1 November 2013, 14th Waternet Symposium, Dar es Salaam, Tanzania, 2013, 4 pp,

Bastiaanssen, W. G. M., Menenti, M., Feddes, R. A., and Holtslag, A. A. M.: A remote sensing surface energy balance algorithm for land (SEBAL). 1. Formulation, Journal of Hydrology, 212–213, 198-212, doi: http://dx.doi.org/10.1016/S0022-1694(98)00253-4, 1998a.

Bastiaanssen, W. G. M., Pelgrum, H., Wang, J., Ma, Y., Moreno, J. F., Roerink, G. J., and van der Wal, T.: A remote sensing surface energy balance algorithm for land (SEBAL).: Part 2: Validation, Journal of Hydrology, 212–213, 213-229, doi: http://dx.doi.org/10.1016/S0022-1694(98)00254-6, 1998b.

Beckers, J., Smerdon, B., and Wilson, M.: Review of hydrologic models for forest management and climate change applications in British Columbia and Alberta., Forrex Forum for Research and Extension in Natural Resources., Kamloops, British Columbia, Canada., 2009.

Belo-Pereira, M., Dutra, E., and Viterbo, P.: Evaluation of global precipitation data sets over the Iberian Peninsula, J. Geophys. Res., 116, D20101, doi:10.1029/2010JD015481, 2011.

Bergström, S.: The HBV model: Its structure and applications, SMHI, RH, 4, Norrköping, Sweden, 1992.

Bergström, S., and Graham, L. P.: On the scale problem in hydrological modelling, Journal of Hydrology, 211, 253-265, 1998.

Bessems, I., Verschuren, D., Russell, J. M., Hus, J., Mees, F., and Cumming, B. F.: Palaeolimnological evidence for widespread late 18th century drought across equatorial East Africa, Palaeogeography, Palaeoclimatology, Palaeoecology, 259, 107-120, 2008.

Beven, K. J., and Kirkby, M. J.: A physically based, variable contributing area model of basin hydrology, Hydrological Sciences Bulletin, 24, 43-69, doi: 10.1080/02626667909491834, 1979.

Bierkens, M., Finke, P., and De Willigen, P.: Upscaling and downscaling methods for environmental research, Kluwer Academic Publishers, Dordrecht, 190 pp., 2000.

Bierkens, M. F. P., Bell, V. A., Burek, P., Chaney, N., Condon, L. E., David, C. H., de Roo, A., Döll, P., Drost, N., Famiglietti, J. S., Flörke, M., Gochis, D. J., Houser, P., Hut, R., Keune, J., Kollet, S., Maxwell, R. M., Reager, J. T., Samaniego, L., Sudicky, E., Sutanudjaja, E. H., van de Giesen, N., Winsemius, H., and Wood, E. F.: Hyper-resolution global hydrological modelling: what is next?, Hydrological Processes, 29, 310-320, doi:10.1002/hyp.10391, 2015.

Blöschl, G., and Sivapalan, M.: Scale issues in hydrological modelling: a review, Hydrological Processes, 9, 251-290, doi: 10.1002/hyp.3360090305, 1995.

Blöschl, G.: Scaling in hydrology, Hydrological Processes, 15, 709-711, doi: 10.1002/hyp.432, 2001.

Blöschl, G.: Statistical Upscaling and Downscaling in Hydrology, in: Encyclopedia of Hydrological Sciences, edited by: Anderson, M. G., John Wiley & Sons, Ltd, 135-154, 2006.

Blöschl, G., Komma, J., and Hasenauer, S.: Hydrological downscaling of soil moisture. Final report of the Visiting Scientist Activity to the Satellite Application Facility on Support to Operational Hydrology and Water Management (H-SAF). , http://hsaf.meteoam.it/documents/reference/HSAF_VS_38_TUWIEN-final-report.pdf, last access: 28 August 2014, 1-64, 2009.

Boussetta, S., Balsamo, G., Beljaars, A., Kral, T., and Jarlan, L.: Impact of a satellite-derived Leaf Area Index monthly climatology in a global Numerical Weather Prediction model, Int. J. Remote Sensing (accepted), also available as ECMWF Tech. Memo. 640, 2011.

Boussetta, S., Balsamo, G., Beljaars, A., Kral, T., and Jarlan, L.: Impact of a satellite-derived leaf area index monthly climatology in a global numerical weather prediction model, Anglais, 34, 3520-3542, doi: 10.1080/01431161.2012.716543, 2012.

Brakhage, D., The Positive Effects of Drought: http://www.ducks.org/conservation/wetlands/the-positive-effects-of-drought, access: 20 November 2014, 2008.

Brown, C., Baroang, K. M., Conrad, E., Lyon, B., Watkins, D., Fiondella, F., Kaheil, Y., Robertson, A., Rodriguez, J., Sheremata, M., and Ward, M. N.: Managing climate risk in water supply systems, IRI technical report 10-15, International Research Institute for Climate and Society, Palisades, NY, available online at: http://iri.columbia.edu/resources/publications/pub_id/1048/, last access: 7 February 2014, 133, 2010.

Busch, F. A., Niemann, J. D., and Coleman, M.: Evaluation of an empirical orthogonal function–based method to downscale soil moisture patterns based on topographical attributes, Hydrological Processes, 26, 2696-2709, doi: 10.1002/hyp.8363, 2012.

Caminade, C., and Terray, L.: Twentieth century Sahel rainfall variability as simulated by the ARPEGE AGCM, and future changes, Climate Dynamics, 35, 75-94, doi: 10.1007/s00382-009-0545-4, 2010.

CEH The Centre for Ecology & Hydrology, GWAVA: Global Water Availability Assessment model: http://www.ceh.ac.uk/sci_programmes/Water/GWAVA.html, access: May 2011, 2011.

CGIAR-CSI Global Aridity and PET Database: http://www.cgiar-csi.org, access: May 2012, 2010.

Chen, J., Brissette, F. P., Chaumont, D., and Braun, M.: Performance and uncertainty evaluation of empirical downscaling methods in quantifying the climate change impacts on hydrology over two North American river basins, Journal of Hydrology, 479, 200-214, doi: http://dx.doi.org/10.1016/j.jhydrol.2012.11.062, 2013.

Cleugh, H. A., Leuning, R., Mu, Q., and Running, S. W.: Regional evaporation estimates from flux tower and MODIS satellite data, Remote Sensing of Environment, 106, 285-304, 2007.

Cloke, H. L., and Pappenberger, F.: Evaluating forecasts of extreme events for hydrological applications: an approach for screening unfamiliar performance measures, Meteorological Applications, 15, 181-197, doi: 10.1002/met.58, 2008.

Coumou, D., and Rahmstorf, S.: A decade of weather extremes, Nature Clim. Change, 2, 491-496, doi: 10.1038/nclimate1452, 2012.

d'Orgeval, T., Polcher, J., and de Rosnay, P.: Sensitivity of the West African hydrological cycle in ORCHIDEE to infiltration processes, Hydrol. Earth Syst. Sci., 12, 1387-1401, 2008.

Dai, A.: Drought under global warming: a review, Wiley Interdisciplinary Reviews: Climate Change, 2, 45-65, doi: 10.1002/wcc.81, 2011.

Dai, A.: Increasing drought under global warming in observations and models, Nature Clim. Change, 3, 52-58, doi: http://www.nature.com/nclimate/journal/v3/n1/abs/nclimate1633.html#supplementary-information, 2013.

Day, G.: Extended Streamflow Forecasting Using NWSRFS, Journal of Water Resources Planning and Management, 111, 157-170, 1985.

De Roo, A. P. J., Wesseling, C. G., and Van Deursen, W. P. A.: Physically based river basin modelling within a GIS: the LISFLOOD model, Hydrological Processes, 14, 1981-1992, doi: 10.1002/1099-1085(20000815/30)14:11/12<1981::AID-HYP49>3.0.CO;2-F, 2000.

Dee, D. P., Uppala, S. M., Simmons, A. J., Berrisford, P., Poli, P., Kobayashi, S., Andrae, U., Balmaseda, M. A., Balsamo, G., Bauer, P., Bechtold, P., Beljaars, A. C. M., van de Berg, L., Bidlot, J., Bormann, N., Delsol, C., Dragani, R., Fuentes, M., Geer, A. J., Haimberger, L., Healy, S. B., Hersbach, H., Hólm, E. V., Isaksen, L., Kållberg, P., Köhler, M., Matricardi, M., McNally, A. P., Monge-Sanz, B. M., Morcrette, J. J., Park, B. K., Peubey, C., de Rosnay, P., Tavolato, C., Thépaut, J. N., and Vitart, F.: The ERA-Interim reanalysis: configuration and performance of the data assimilation system, Quarterly Journal of the Royal Meteorological Society, 137, 553-597, doi: 10.1002/qj.828, 2011.

Department of Agriculture of South Africa: Crops and markets - First quarter 2006, Vol 87., No. 927, Directorate Agricultural Statistics - Department of Agriculture http://www.daff.gov.za/docs/statsinfo/Crops_0106.pdf, last access: December 2013, 2006.

DEWFORA: WP2-D2.2 - Inventory of institutial frameworks and drought mitigation and adaptation practices in Africa, DEWFORA project - EU FP7, www.dewfora.net, last access: August 2014, 2011.

DEWFORA: WP6-D6.1 - Implementation of improved methodologies in comparative case studies - Inception report for each case study, DEWFORA Project - EU FP7, www.dewfora.net, last access: December 2013, 2012a.

DEWFORA: WP5-D5.1 - Concept report describing the outline of a framework for drought warning and mitigation in Africa DEWFORA project - EU FP7 www.dewfora.net, last access: July 2014, 2012b.

DEWFORA: Science Policy Brief (Africa) - Implementing drought early warning systems in Africa: policy lessons and future needs, http://www.unesco-ihe.org/sites/default/files/dewfora_policy_brief_africa_-_digital_version.pdf, access: 06/12/2014, 2013a.

DEWFORA: WP6-D6.2 - Limpopo Case Study, Application of the DEWFORA Drought Early Warning Framework DEWFORA project - EU FP7 www.dewfora.net, last access: July 2014, 2013b.

Dingman, S. L.: Physical hydrology, Prentice Hall Englewood Cliffs, NJ, 1994.

Döll, P., Kaspar, F., and Lehner, B.: A global hydrological model for deriving water availability indicators: model tuning and validation, Journal of Hydrology, 270, 105-134, 2003.

Douville, H., Viterbo, P., Mahfouf, J. F., and Beljaars, A. C. M.: Evaluation of the Optimum interpolation and nudging techniques for soil moisture analysis using FIFE data, Mon. Weather Rev., 128, 1733 –1756 2000.

Droogers, P., and Allen, R.: Estimating Reference Evapotranspiration Under Inaccurate Data Conditions, Irrigation and Drainage Systems, 16, 33-45, 2002.

Droogers, P., Immerzeel, W. W., Terink, W., Hoogeveen, J., Bierkens, M. F. P., van Beek, L. P. H., and Debele, B.: Water resources trends in Middle East and North Africa towards 2050, Hydrol. Earth Syst. Sci., 16, 3101-3114, doi: 10.5194/hess-16-3101-2012, 2012.

Drusch, M., Scipal, K., de Rosnay, P., Balsamo, G., Andersson, E., Bougeault, P., and Viterbo, P.: Exploitation of satellite data in the surface analysis. ECMWF Tech. Memo. No. 576, http://www.ecmwf.int/publications/library/do/references/show?id=88712, 2008.

Druyan, L. M.: Studies of 21st-century precipitation trends over West Africa, International Journal of Climatology, 31, 1415-1424, doi: 10.1002/joc.2180, 2011.

Dube, O. P., and Sekhwela, M. B. M.: Community coping strategies in Semiarid Limpopo basin part of Botswana: Enhancing adaptation capacity to climate change, http://www.aiaccproject.org/working_papers/Working%20Papers/AIACC_WP47_Dube.pdf, last access: December, 2013, 1-40, 2007.

Dürr, H. H., Meybeck, M., and Dürr, S. H.: Lithologic composition of the Earth's continental surfaces derived from a new digital map emphasizing riverine material transfer, Global Biogeochemical Cycles, 19, GB4S10, doi: 10.1029/2005GB002515, 2005.

Dutra, E., Di Giuseppe, F., Wetterhall, F., and Pappenberger, F.: Seasonal forecasts of droughts in African basins using the Standardized Precipitation Index, Hydrol. Earth Syst. Sci., 17, 2359-2373, doi:10.5194/hess-17-2359-2013, 2013a.

Dutra, E., Magnusson, L., Wetterhall, F., Cloke, H. L., Balsamo, G., Boussetta, S., and Pappenberger, F.: The 2010–2011 drought in the Horn of Africa in ECMWF reanalysis and seasonal forecast products, International Journal of Climatology, 33, 1720-1729, doi: 10.1002/joc.3545, 2013b.

Dutra, E., Pozzi, W., Wetterhall, F., Di Giuseppe, F., Magnusson, L., Naumann, G., Barbosa, P., Vogt, J., and Pappenberger, F.: Global meteorological drought - Part 2: Seasonal forecasts, Hydrol. Earth Syst. Sci. , 18, 2669-2678, doi:10.5194/hess-18-2669-2014, 2014.

DWA: Tzaneen Dam, Department of Water Affairs, South Africa, 2013.

Ek, M., Xia, Y., and the NLDAS team: NCEP/EMC NLDAS Support for Drought Monitoring and Seasonal Prediction, US National Oceanic and Atmospheric Administration, Climate Test Bed Joint Seminar Series, NASA, Goddard Visitor Center, Greenbelt, Maryland, http://www.nws.noaa.gov/ost/climate/STIP/FY10CTBSeminars/mek_041410.pdf, access 01/12/2014, 2010.

Elagib, N. A., and Elhag, M. M.: Major climate indicators of ongoing drought in Sudan, Journal of Hydrology, 409, 612-625, doi: http://dx.doi.org/10.1016/j.jhydrol.2011.08.047, 2011.

EM-DAT: The OFDA/CRED International Disaster Database, Universite catholique de Louvain, Brussels, Belgium, Created on:Mar-27-2014. - Data version: v12.07, access: 27 March 2014, 2014.

Endfield, G. H., and Nash, D. J.: Drought, desiccation and discourse: missionary correspondence and nineteenth-century climate change in central southern Africa, The Geographical Journal, 168, 33-47, doi: 10.1111/1475-4959.00036, 2002.

EUMETSAT: The EUMETSAT Satellite Application Facility on Land Surface Analysis (LSA SAF) - Product User Manual, Evapotranspiration (ET), 2011.

FAO: Irrigation Potential in Africa: A Basin Approach, FAO-UN, Rome, 1997.

FAO, The digital soil map of the world (Version 3.6): http://www.fao.org/geonetwork/srv/en/metadata.show?id=14116&currTab=distribution, access: 21 August 2012, 2003.

FAO: Drought impact mitigation and prevention in the Limpopo River Basin - A situation analysis, FAO, Rome, http://www.fao.org/docrep/008/y5744e/y5744e00.htm#Contents, access: 03/12/2014, 2004.

Farmer, C. L.: Upscaling: a review, International Journal for Numerical Methods in Fluids, 40, 63-78, doi: 10.1002/fld.267, 2002.

Fekete, B. M., Vörösmarty, C. J., Roads, J. O., and Willmott, C. J.: Uncertainties in precipitation and their impacts on runoff estimates, Journal of Climate, 17, 294-304, 2004.

Ferrer, J., Pérez, M., Pérez, F., and Artés, J.: Specific combined actions in Turia River during 2005-2007 drought, Options Méditerranéennes, Series A, No.80: Séminaires Méditerranéens (CIHEAM), http://www.iamz.ciheam.org/medroplan/a-80_OPTIONS/Sesion%202/%28227-234%29%2034%20Ferrer-Pe-Pe%20PS2.pdf, access: 5 December 2014, 227-234, 2008.

Fowler, H. J., Blenkinsop, S., and Tebaldi, C.: Linking climate change modelling to impacts studies: recent advances in downscaling techniques for hydrological modelling, International Journal of Climatology, 27, 1547-1578, doi: 10.1002/joc.1556, 2007.

Friederichs, P., and Thorarinsdottir, T. L.: Forecast verification for extreme value distributions with an application to probabilistic peak wind prediction, Environmetrics, 23, 579-594, 2012.

García, M., Sandholt, I., Ceccato, P., Mougin, E., Kergoat, L., and Timouk, F.: Estimating evaportranspiration in the Sahel using remote sensing products 30th AMS Conference on Agricultural and Forest Meteorology, Boston, USA, 2012.

Gerten, D., Schaphoff, S., Haberlandt, U., Lucht, W., and Sitch, S.: Terrestrial vegetation and water balance--hydrological evaluation of a dynamic global vegetation model, Journal of Hydrology, 286, 249-270, 2004.

Gharari, S., Hrachowitz, M., Fenicia, F., and Savenije, H. H. G.: Hydrological landscape classification: investigating the performance of HAND based landscape classifications in a central European meso-scale catchment, Hydrol. Earth Syst. Sci., 15, 3275-3291, doi: 10.5194/hess-15-3275-2011, 2011.

Ghilain, N., Arboleda, A., and Gellens-Meulenberghs, F.: Evapotranspiration modelling at large scale using near-real time MSG SEVIRI derived data, Hydrol. Earth Syst. Sci., 15, 771-786, doi: 10.5194/hess-15-771-2011, 2011.

Giannini, A., Saravanan, R., and Chang, P.: Oceanic Forcing of Sahel Rainfall on Interannual to Interdecadal Time Scales, Science, 302, 1027-1030, 2003.

Giannini, A., Biasutti, M., and Verstraete, M. M.: A climate model-based review of drought in the Sahel: Desertification, the re-greening and climate change, Global and Planetary Change, 64, 119-128, doi: http://dx.doi.org/10.1016/j.gloplacha.2008.05.004, 2008.

Glantz, M. H. e.: Drought and Hunger in Africa: Denying Famine a Future, Cambridge University Press, Cambridge, 1987.

Gosling, S. N., and Arnell, N. W.: Simulating current global river runoff with a global hydrological model: model revisions, validation, and sensitivity analysis, Hydrological Processes, 25, 1129-1145, 2010.

Grasso, V. F.: Early Warning Systems: State-of-Art Analysis and Future Directions - Draft Report, Division of Early Warning and Assessment (DEWA), United Nations Environment Programme (UNEP), available at: https://na.unep.net/geas/docs/Early_Warning_System_Report.pdf, last access: 28 March 2015, Nairobi, 66 pp, 2009.

Green, T. R., and Erskine, R. H.: Measurement, scaling, and topographic analyses of spatial crop yield and soil water content, Hydrological Processes, 18, 1447-1465, doi: 10.1002/hyp.1422, 2004.

Guttman, N. B.: Comparing the Palmer Drought Index and the Standardized Precipitation Index, JAWRA Journal of the American Water Resources Association, 34, 113-121, doi: 10.1111/j.1752-1688.1998.tb05964.x, 1998.

Haddeland, I., Clark, D. B., Franssen, W., Ludwig, F., Voß, F., Arnell, N. W., Bertrand, N., Best, M., Folwell, S., Gerten, D., Gomes, S., Gosling, S. N., Hagemann, S., Hanasaki, N., Harding, R., Heinke, J., Kabat, P., Koirala, S., Oki, T., Polcher, J., Stacke, T., Viterbo, P., Weedon, G. P., and Yeh., P.: Multi-Model Estimate of the Global Terrestrial Water Balance: Setup and First Results, Journal of Hydrometeorology, 12, 2011.

Hagemann, S., Botzet, M., Dümenil, L., and Machenhauer, B.: Derivation of global GCM boundary conditions from 1 km land use satellite data, MPI Report No. 289, Max Planck Institute for Meteorology, Hamburg, 1999.

Hagemann, S.: An improved land surface parameter dataset for global and regional climate models, MPI Report No. 336, Max Planck Institute for Meteorology, Hamburg, 2002.

Hamlet, A., and Lettenmaier, D.: Columbia River Streamflow Forecasting Based on ENSO and PDO Climate Signals, Journal of Water Resources Planning and Management, 125, 333-341, doi:10.1061/(ASCE)0733-9496(1999)125:6(333), 1999.

Hansen, J., Sato, M., and Ruedy, R.: Perception of climate change, Proceedings of the National Academy of Sciences, 109, E2415-E2423, doi:10.1073/pnas.1205276109, 2012.

Hargreaves, G. H., and Allen, R. G.: History and Evaluation of Hargreaves Evapotranspiration Equation, Journal of Irrigation and Drainage Engineering, 129, 53-63, 2003.

Hastenrath, S., Polzin, D., and Mutai, C.: Diagnosing the 2005 Drought in Equatorial East Africa, Journal of Climate, 20, 4628-4637, doi: 10.1175/jcli4238.1, 2007.

Heim, R. R.: A Review of Twentieth-Century Drought Indices Used in the United States, Bulletin of the American Meteorological Society, 83, 1149-1165, doi:10.1175/1520-0477(2002)083<1149:AROTDI>2.3.CO;2, 2002.

Herweijer, C., and Seager, R.: The global footprint of persistent extra-tropical drought in the instrumental era, International Journal of Climatology, 28, 1761-1774, doi: 10.1002/joc.1590, 2008.

Hewitson, B. C., and Crane, R. G.: Climate downscaling: techniques and application, Climate Research, 07, 85-95, doi: 10.3354/cr0007085, 1996.

Hijmans, R. J., Cameron, S. E., Parra, J. L., Jones, P. G., and Jarvis, A.: Very high resolution interpolated climate surfaces for global land areas, International Journal of Climatology, 25, 1965-1978, 2005.

Hirabayashi, Y., Kanae, S., Struthers, I., and Oki, T.: A 100-year (1901–2000) global retrospective estimation of the terrestrial water cycle, Journal of Geophysical Research, 110, D19101, 2005.

HMNDP: High Level Meeting on National Drought Policy (HMNDP) - Final Declaration, http://www.hmndp.org/sites/default/files/docs/HMNDP_Final_Declaration.pdf, access: 05/12/2014, 2013.

Huffman, G. J., Adler, R. F., Bolvin, D. T., and Gu, G.: Improving the global precipitation record: GPCP version 2.1, Geophys. Res. Lett, 36, L17808, doi:10.1029/2009GL040000, 2009.

Huntington, T. G.: Evidence for intensification of the global water cycle: Review and synthesis, Journal of Hydrology, 319, 83-95, doi: http://dx.doi.org/10.1016/j.jhydrol.2005.07.003, 2006.

Hwang, Y.-T., Frierson, D. M. W., and Kang, S. M.: Anthropogenic sulfate aerosol and the southward shift of tropical precipitation in the late 20th century, Geophysical research letters, 40, 2845-2850, doi: 10.1002/grl.50502, 2013.

IPCC: Summary for Policymakers. In: Climate Change 2007: The Physical Science Basis. Contribution of Working Group I to the Fourth Assessment Report of the Intergovernmental Panel on Climate Change [Solomon, S., D. Qin, M. Manning, Z. Chen, M. Marquis, K.B. Averyt, M.Tignor and H.L.

Miller (eds.)]. Cambridge University Press, Cambridge, United Kingdom and New York, NY, USA, available at: https://www.ipcc.ch/pdf/assessment-report/ar4/wg1/ar4-wg1-spm.pdf, last access: 28 March 2015, 1-18, 2007a.

IPCC: Climate Change 2007: The Physical Science Basis. Contribution of Working Group I to the Fourth Assessment, Report of the Intergovernmental Panel on Climate Change [Solomon, S., D. Qin, M. Manning, Z. Chen, M. Marquis, K.B. Averyt, M. Tignor and H.L. Miller (eds.)], Cambridge University Press, Cambridge, United Kingdom and New York, NY, USA, 996 pp., 2007b.

Jacovides, C. P., Tymvios, F. S., Asimakopoulos, D. N., Theofilou, K. M., and Pashiardes, S.: Global photosynthetically active radiation and its relationship with global solar radiation in the Eastern Mediterranean basin, Theor. Appl. Climatol., 74, 227-233, 2003.

Jakob Themeßl, M., Gobiet, A., and Leuprecht, A.: Empirical-statistical downscaling and error correction of daily precipitation from regional climate models, International Journal of Climatology, 31, 1530-1544, doi: 10.1002/joc.2168, 2011.

Jiménez, C., Prigent, C., Mueller, B., Seneviratne, S. I., McCabe, M. F., Wood, E. F., Rossow, W. B., Balsamo, G., Betts, A. K., Dirmeyer, P. A., Fisher, J. B., Jung, M., Kanamitsu, M., Reichle, R. H., Reichstein, M., Rodell, M., Sheffield, J., Tu, K., and Wang, K.: Global intercomparison of 12 land surface heat flux estimates, Journal of Geophysical Research: Atmospheres, 116, D02102, doi: 10.1029/2010jd014545, 2011.

JRC European Commission - Institute for Environment and Sustainability, LISFLOOD Model: http://floods.jrc.ec.europa.eu/lisflood-model, access: 1 July 2011, 2011.

Kang, T.-H., Kim, Y.-O., and Hong, I.-P.: Comparison of pre- and post-processors for ensemble streamflow prediction, Atmospheric Science Letters, 11, 153-159, doi: 10.1002/asl.276, 2010.

Kasei, R., Diekkrüger, B., and Leemhuis, C.: Drought frequency in the Volta Basin of West Africa, Sustain Sci, 5, 89-97, doi: 10.1007/s11625-009-0101-5, 2010.

Keyantash, J., and Dracup, J. A.: The quantification of drought: an evaluation of drought indices, Bulletin of the American Meteorological Society, 83, 1167-1180, 2002.

Keys, R.: Cubic convolution interpolation for digital image processing, Acoustics, Speech and Signal Processing, IEEE Transactions on, 29, 1153-1160, 1981.

Kim, H. W., Hwang, K., Mu, Q., Lee, S. O., and Choi, M.: Validation of MODIS 16 global terrestrial evapotranspiration products in various climates and land cover types in Asia, KSCE Journal of Civil Engineering, 16, 229-238, 2012.

Kingston, D. G., Todd, M. C., Taylor, R. G., Thompson, J. R., and Arnell, N. W.: Uncertainty in the estimation of potential evapotranspiration under climate change, Geophys. Res. Lett., 36, L20403, doi: 10.1029/2009GL040267, 2009.

Kirtman, B. P., Min, D., Infanti, J. M., Kinter, J. L., Paolino, D. A., Zhang, Q., van den Dool, H., Saha, S., Mendez, M. P., Becker, E., Peng, P., Tripp, P., Huang, J., DeWitt, D. G., Tippett, M. K., Barnston, A. G., Li, S., Rosati, A., Schubert, S. D., Rienecker, M., Suarez, M., Li, Z. E., Marshak, J., Lim, Y.-K., Tribbia, J., Pegion, K., Merryfield, W. J., Denis, B., and Wood, E. F.: The North American Multimodel Ensemble: Phase-1 Seasonal-to-Interannual Prediction; Phase-2 toward Developing

Intraseasonal Prediction, Bulletin of the American Meteorological Society, 95, 585-601, doi: 10.1175/bams-d-12-00050.1, 2013.

Kling, H., Fuchs, M., and Paulin, M.: Runoff conditions in the upper Danube basin under an ensemble of climate change scenarios, Journal of Hydrology, 424–425, 264-277, doi: http://dx.doi.org/10.1016/j.jhydrol.2012.01.011, 2012.

Krysanova, V., Müller-Wohlfeil, D.-I., and Becker, A.: Development and test of a spatially distributed hydrological/water quality model for mesoscale watersheds, Ecological Modelling, 106, 261-289, 1998.

Krysanova, V., and Wechsung, F.: SWIM (Soil and Water Integrated Model) User Manual, Potsdam Institute for Climate Impact Research, Potsdam, Germany, 2000.

Ladson, A. R., and Moore, I. D.: Soil water prediction on the Konza Prairie by microwave remote sensing and topographic attributes, Journal of Hydrology, 138, 385-407, doi: http://dx.doi.org/10.1016/0022-1694(92)90127-H, 1992.

Landman, W. A., and Mason, S. J.: Operational long-lead prediction of South African rainfall using canonical correlation analysis, International Journal of Climatology, 19, 1073-1090, doi: 10.1002/(sici)1097-0088(199908)19:10<1073::aid-joc415>3.0.co;2-j, 1999.

Landman, W. A., Mason, S. J., Tyson, P. D., and Tennant, W. J.: Statistical downscaling of GCM simulations to Streamflow, Journal of Hydrology, 252, 221-236, doi: http://dx.doi.org/10.1016/S0022-1694(01)00457-7, 2001.

Lau, K. M., and Wu, H. T.: Warm rain processes over tropical oceans and climate implications, Geophysical research letters, 30, 2290, doi: 10.1029/2003GL018567, 2003.

LBPTC: Joint Limpopo River Basin Study Scoping Phase. Final Report. BIGCON Consortium., Limpopo Basin Permanent Technical Committee, http://www.limcom.org/_system/writable/DMSStorage/1031en/LIMCOM2010_ScopingStudy_Eng .pdf, access: 26 August 2014, 2010.

Lebel, T., Cappelaere, B., Galle, S., Hanan, N., Kergoat, L., Levis, S., Vieux, B., Descroix, L., Gosset, M., Mougin, E., Peugeot, C., and Seguis, L.: AMMA-CATCH studies in the Sahelian region of West-Africa: An overview, Journal of Hydrology, 375, 3-13, doi: http://dx.doi.org/10.1016/j.jhydrol.2009.03.020, 2009.

Lehner, B., and Döll, P.: Global Lakes and Wetlands Database GLWD, GLWD Docu mentation, 2004.

Lehner, B., Döll, P., Alcamo, J., Henrichs, T., and Kaspar, F.: Estimating the impact of global change on flood and drought risks in Europe: a continental, integrated analysis, Climatic Change, 75, 273-299, doi: 10.1007/s10584-006-6338-4, 2006.

Liang, X., Lettenmaier, D. P., Wood, E. F., and Burges, S. J.: A simple hydrologically based model of land surface water and energy fluxes for general circulation models, J. Geophys. Res, 99, 415-414, doi: 10.1029/94JD00483, 1994.

Lin, C.: Soil moisture analysis and seasonal forecast of drought - Final Progress Report, Drought Research Initiative (DRI), Canada, 2010.

Lindström, G., Johansson, B., Persson, M., Gardelin, M., and Bergström, S.: Development and test of the distributed HBV-96 hydrological model, Journal of Hydrology, 201, 272-288, 1997.

Liu, Y., Yamaguchi, Y., and Ke, C.: Reducing the discrepancy between ASTER and MODIS land surface temperature products, Sensors, 7, 3043-3057, 2007.

Lloyd-Hughes, B., and Saunders, M. A.: A drought climatology for Europe, International Journal of Climatology, 22, 1571-1592, doi: 10.1002/joc.846, 2002.

Loh, W.-Y.: Classification and regression trees, Wiley Interdisciplinary Reviews: Data Mining and Knowledge Discovery, 1, 14-23, doi: 10.1002/widm.8, 2011.

Lohmann, D., Lettenmaier, D. P., Liang, X., Wood, E. F., Boone, A., Chang, S., Chen, F., Dai, Y., Desborough, C., Dickinson, R. E., Duan, Q., Ek, M., Gusev, Y. M., Habets, F., Irannejad, P., Koster, R., Mitchell, K. E., Nasonova, O. N., Noilhan, J., Schaake, J., Schlosser, A., Shao, Y., Shmakin, A. B., Verseghy, D., Warrach, K., Wetzel, P., Xue, Y., Yang, Z.-L., and Zeng, Q.-c.: The Project for Intercomparison of Land-surface Parameterization Schemes (PILPS) phase 2(c) Red-Arkansas River basin experiment:: 3. Spatial and temporal analysis of water fluxes, Global and Planetary Change, 19, 161-179, 1998.

Loon, A. F. v., Lanen, H. A. J. v., Seibert, J., and Torfs, P. J. J. F.: Adaptation of the HBV model for the study of drought propagation in European catchments, EGU General Assembly Vienna, Austria, 19-24 April 2009, 9589-9589, 2009,

Lott, F. C., Christidis, N., and Stott, P. A.: Can the 2011 East African drought be attributed to human-induced climate change?, Geophysical research letters, 40, 1177-1181, doi: 10.1002/grl.50235, 2013.

Love, D., Uhlenbrook, S., and van der Zaag, P.: Challenges of regionalising a conceptual model in semi-arid meso-catchments, 10th International WATERNET/WARFSA/GWP-SA Symposium, Entebe, Uganda, 23, 2009,

Loveland, T. R., Reed, B. C., Brown, J. F., Ohlen, D. O., Zhu, Z., Yang, L., and Merchant, J. W.: Development of a global land cover characteristics database and IGBP DISCover from 1 km AVHRR data, Journal of Remote Sensing, 21, 1303--1330, 2000.

Manatsa, D., Chingombe, W., Matsikwa, H., and Matarira, C. H.: The superior influence of Darwin Sea level pressure anomalies over ENSO as a simple drought predictor for Southern Africa, Theor. Appl. Climatol., 92, 1-14, doi: 10.1007/s00704-007-0315-3, 2008.

Martina, M. L. V., and Todini, E.: Watershed Hydrological Modeling: Toward Physically Meaningful Processes Representation, in: Hydrological Modelling and the Water Cycle, edited by: Sorooshian, S., Hsu, K.-L., Coppola, E., Tomassetti, B., Verdecchia, M., and Visconti, G., Water Science and Technology Library, Springer Berlin Heidelberg, 229-241, 2008.

Masih, I., Maskey, S., Mussá, F. E. F., and Trambauer, P.: A review of droughts on the African continent: a geospatial and long-term perspective, Hydrol. Earth Syst. Sci., 18, 3635-3649, doi: 10.5194/hess-18-3635-2014, 2014.

Maskey, S., and Trambauer, P.: Hydrological modeling for drought assessment, in: Hydro-Meteorological Hazards, Risks, and Disasters, 1 ed., edited by: Shroder, J. F., Paron, P., and Di Baldassarre, G., Elsevier, 263-282, 2014.

McClain, M.: Balancing Water Resources Development and Environmental Sustainability in Africa: A Review of Recent Research Findings and Applications, Ambio, 42, 549-565, doi: 10.1007/s13280-012-0359-1, 2013.

McKee, T. B., Doesken, N. J., and Kleist, J.: The relationship of drought frequency and duration to time scales, Proceedings of the 8th Conference on Applied Climatology, Vol. 17. No. 22, Boston, MA: American Meteorological Society, 17, 179-183, 1993.

Meigh, J. R., McKenzie, A. A., and Sene, K. J.: A Grid-Based Approach to Water Scarcity Estimates for Eastern and Southern Africa, Water Resources Management, 13, 85-115, 1999.

Melone, F., Barbetta, S., Diomede, T., Peruccacci, S., Rossi, M., Tessarolo, A., and Verdecchia, M.: Report: Review and selection of hydrological models - Integration of hydrological models and meteorological inputs., RISK-AWARE RISK-Advanced Weather forecast system to Advice on Risk Events and management (http://www.smr.arpa.emr.it/riskaware/), 2005.

Mendicino, G., Senatore, A., and Versace, P.: A Groundwater Resource Index (GRI) for drought monitoring and forecasting in a mediterranean climate, Journal of Hydrology, 357, 282-302, doi: http://dx.doi.org/10.1016/j.jhydrol.2008.05.005, 2008.

Milly, P. C. D., and Shmakin, A. B.: Global Modeling of Land Water and Energy Balances. Part I: The Land Dynamics (LaD) Model, Journal of Hydrometeorology, 3, 283-299, 2002.

Miralles, D. G., Gash, J. H., Holmes, T. R. H., de Jeu, R. A. M., and Dolman, A. J.: Global canopy interception from satellite observations, Journal of Geophysical Research: Atmospheres, 115, D16122, doi: 10.1029/2009jd013530, 2010.

Miralles, D. G., De Jeu, R. A. M., Gash, J. H., Holmes, T. R. H., and Dolman, A. J.: Magnitude and variability of land evaporation and its components at the global scale, Hydrol. Earth Syst. Sci., 15, 967-981, 2011a.

Miralles, D. G., Holmes, T. R. H., De Jeu, R. A. M., Gash, J. H., Meesters, A. G. C. A., and Dolman, A. J.: Global land-surface evaporation estimated from satellite-based observations, Hydrol. Earth Syst. Sci., 15, 453-469, 2011b.

Miralles, D. G., van den Berg, M. J., Gash, J. H., Parinussa, R. M., de Jeu, R. A. M., Beck, H. E., Holmes, T. R. H., Jiménez, C., Verhoest, N. E. C., Dorigo, W. A., Teuling, A. J., and Johannes Dolman, A.: El Niño–La Niña cycle and recent trends in continental evaporation, Nature Clim. Change, 4, 122–126, doi:10.1038/nclimate2068, http://www.nature.com/nclimate/journal/vaop/ncurrent/abs/nclimate2068.html#supplementary-information, 2014.

Mishra, A. K., and Singh, V. P.: A review of drought concepts, Journal of Hydrology, 391, 202-216, 2010.

Mishra, V., Cherkauer, K. A., and Shukla, S.: Assessment of Drought due to Historic Climate Variability and Projected Future Climate Change in the Midwestern United States, Journal of Hydrometeorology, 11, 46-68, 2010.

Molteni, F., Stockdale, T., Balmaseda, M. A., BALSAMO, G., Buizza, R., Ferranti, L., Magnunson, L., Mogensen, K., Palmer, T., and Vitart, F.: The new ECMWF seasonal forecast system (System 4), ECMWF Tech. Memo., 656, 49 pp., 2011.

Moore, I., Burch, G., and Mackenzie, D.: Topographic effects on the distribution of surface soil water and the location of ephemeral gullies, Trans. Am. Soc. Agric. Engin. (USA), 31, 1098-1107, doi: 10.13031/2013.30829, 1988.

Moore, I. D., Grayson, R. B., and Ladson, A. R.: Digital terrain modelling: A review of hydrological, geomorphological, and biological applications, Hydrological Processes, 5, 3-30, doi: 10.1002/hyp.3360050103, 1991.

Moore, R.: The probability-distributed principle and runoff production at point and basin scales/Le principe de la distribution des probabilités et la production d'écoulement en un point et à l'échelle d'un bassin, Hydrological Sciences Journal, 30, 273-297, 1985.

Morgan, R.: The development and applications of a Drought Early Warning System in Botswana, Disasters, 9, 44-50, doi: 10.1111/j.1467-7717.1985.tb00909.x, 1985.

Moriasi, D., Arnold, J., Van Liew, M., Bingner, R., Harmel, R., and Veith, T.: Model evaluation guidelines for systematic quantification of accuracy in watershed simulations, http://swat.tamu.edu/media/1312/moriasimodeleval.pdf, Transactions of the ASABE, 50, 885-900, 2007.

Motha, R. P.: Use of Crop Models for Drought Analysis, in: Agricultural Drought Indices. Proceedings of the WMO/UNISDR Expert Group Meeting on Agricultural Drought Indices, 2-4 June 2010, Murcia, Spain, edited by: Sivakumar, M. V. K., Motha, R. P., Wilhite, D. A., and Wood, D. A., World Meteorological Organization. AGM-11, WMO/TD No. 1572; WAOB-2011, Geneva, Switzerland, 197, 2011.

Mu, Q., Heinsch, F. A., Zhao, M., and Running, S. W.: Development of a global evapotranspiration algorithm based on MODIS and global meteorology data, Remote Sensing of Environment, 111, 519-536, 2007.

Mu, Q., Zhao, M., and Running, S. W.: Improvements to a MODIS global terrestrial evapotranspiration algorithm, Remote Sensing of Environment, 115, 1781-1800, 2011.

Mueller, B., Seneviratne, S. I., Jimenez, C., Corti, T., Hirschi, M., Balsamo, G., Ciais, P., Dirmeyer, P., Fisher, J. B., Guo, Z., Jung, M., Maignan, F., McCabe, M. F., Reichle, R., Reichstein, M., Rodell, M., Sheffield, J., Teuling, A. J., Wang, K., Wood, E. F., and Zhang, Y.: Evaluation of global observations-based evapotranspiration datasets and IPCC AR4 simulations, Geophysical research letters, 38, L06402, doi: 10.1029/2010gl046230, 2011.

Mueller, B., Hirschi, M., Jimenez, C., Ciais, P., Dirmeyer, P. A., Dolman, A. J., Fisher, J. B., Jung, M., Ludwig, F., Maignan, F., Miralles, D., McCabe, M. F., Reichstein, M., Sheffield, J., Wang, K. C., Wood, E. F., Zhang, Y., and Seneviratne, S. I.: Benchmark products for land evapotranspiration: LandFlux-EVAL multi-dataset synthesis, Hydrol. Earth Syst. Sci. , 17, 3707-3720, doi:10.5194/hess-17-3707-2013, 2013.

Narasimhan, B., and Srinivasan, R.: Development and evaluation of Soil Moisture Deficit Index (SMDI) and Evapotranspiration Deficit Index (ETDI) for agricultural drought monitoring, Agricultural and Forest Meteorology, 133, 69-88, doi: http://dx.doi.org/10.1016/j.agrformet.2005.07.012, 2005.

Naumann, G., Barbosa, P., Garrote, L., Iglesias, A., and Vogt, J.: Exploring drought vulnerability in Africa: an indicator based analysis to be used in early warning systems, Hydrol. Earth Syst. Sci., 18, 1591-1604, doi: 10.5194/hess-18-1591-2014, 2014.

NDMC National Drought Mitigation Center, Measuring Drought, http://drought.unl.edu/ranchplan/DroughtBasics/WeatherDrought/MeasuringDrought.aspx, access: 13 January 2015, 2015.

Ngo-Duc, T., Polcher, J., and Laval, K.: A 53-year forcing data set for land surface models, J. Geophys. Res., 110, D06116, 2005.

Nicholson, S. E., and Kim, J.: The relationship of the El Niño–Southern Oscillation to African rainfall, International Journal of Climatology, 17, 117-135, doi: 10.1002/(sici)1097-0088(199702)17:2<117::aid-joc84>3.0.co;2-o, 1997.

Nijssen, B., Lettenmaier, D. P., Liang, X., Wetzel, S. W., and Wood, E. F.: Streamflow simulation for continental-scale river basins, Water Resour. Res., 33, 711-724, 1997.

Nijssen, B., O'Donnell, G. M., Lettenmaier, D. P., Lohmann, D., and Wood, E. F.: Predicting the Discharge of Global Rivers, Journal of Climate, 14, 3307-3323, 2001a.

Nijssen, B., Schnur, R., and Lettenmaier, D. P.: Global Retrospective Estimation of Soil Moisture Using the Variable Infiltration Capacity Land Surface Model, Journal of Climate, 14, 1790-1808, 2001b.

NOAA Historical El Nino/ La Nina episodes (1950-present): http://www.cpc.ncep.noaa.gov/products/analysis_monitoring/ensostuff/ensoyears.shtml, access: 31 January 2014, 2014.

Nobre, A. D., Cuartas, L. A., Hodnett, M., Rennó, C. D., Rodrigues, G., Silveira, A., Waterloo, M., and Saleska, S.: Height Above the Nearest Drainage – a hydrologically relevant new terrain model, Journal of Hydrology, 404, 13-29, doi: http://dx.doi.org/10.1016/j.jhydrol.2011.03.051, 2011.

Overgaard, J., Rosbjerg, D., and Butts, M.: Land-surface modelling in hydrological perspective - a review, Biogeosciences, 3, 229-241, 2006.

Owe, M., de Jeu, R., and Holmes, T.: Multisensor historical climatology of satellite-derived global land surface moisture, Journal of Geophysical Research: Earth Surface, 113, F01002, doi:10.1029/2007JF000769, 2008.

Pagano, T. C., Garen, D. C., Perkins, T. R., and Pasteris, P. A.: Daily Updating of Operational Statistical Seasonal Water Supply Forecasts for the western U.S.1, JAWRA Journal of the American Water Resources Association, 45, 767-778, doi: 10.1111/j.1752-1688.2009.00321.x, 2009.

Palmer, W. C.: Meteorological drought, Research paper no.45, US Department of Commerce, Weather Bureau Washington, DC, USA, 1-58, 1965.

Patz, J. A., Campbell-Lendrum, D., Holloway, T., and Foley, J. A.: Impact of regional climate change on human health, Nature, 438, 310-317, 2005.

Peters, E.: Propagation of drought through groundwater systems: illustrated in the Pang (UK) and Upper-Guadiana (ES) catchments, Wageningen Universiteit., 2003.

Peters, E., and Van Lanen, H. A. J.: Propagation of drought in groundwater in semiarid and sub-humid climatic regimes, in: Hydrology in Mediterranean and semiarid regions: International conference, Montpellier, France, 2003, 312–317, IAHS Press, Wallingford, UK, 2003.

Pilgrim, D. H., Chapman, T. G., and Goran, D. G.: Problems of rainfall-runoff modelling in arid and semiarid regions, Hydrological Sciences Journal, 33:4, 379-400, doi: 10.1080/02626668809491261, 1988.

Portmann, F., Siebert, S., Bauer, C., and Döll, P.: Global data set 895 of monthly growing areas of 26 irrigated crops, Frankfurt Hydrology Paper 06, Institute of Physical Geography, University of Frankfurt, Frankfurt am Main, Germany, 2008.

Portmann, F. T., Siebert, S., and Döll, P.: MIRCA2000—Global monthly irrigated and rainfed crop areas around the year 2000: A new high-resolution data set for agricultural and hydrological modeling, Global Biogeochemical Cycles, 24, GB1011, doi: 10.1029/2008gb003435, 2010.

Pozzi, W., Sheffield, J., Stefanski, R., Cripe, D., Pulwarty, R., Vogt, J. V., Heim, R. R., Brewer, M. J., Svoboda, M., Westerhoff, R., van Dijk, A. I. J. M., Lloyd-Hughes, B., Pappenberger, F., Werner, M., Dutra, E., Wetterhall, F., Wagner, W., Schubert, S., Mo, K., Nicholson, M., Bettio, L., Nunez, L., van Beek, R., Bierkens, M., de Goncalves, L. G. G., de Mattos, J. G. Z., and Lawford, R.: Toward Global Drought Early Warning Capability: Expanding International Cooperation for the Development of a Framework for Monitoring and Forecasting, Bulletin of the American Meteorological Society, 94, 776-785, doi: 10.1175/bams-d-11-00176.1, 2013.

Pradhan, N. R., Tachikawa, Y., and Takara, K.: A downscaling method of topographic index distribution for matching the scales of model application and parameter identification, Hydrological Processes, 20, 1385-1405, doi: 10.1002/hyp.6098, 2006.

PROMISE: Predictability and variability of monsoons, and the agricultural and hydrological impacts of climate change, NERC, CEH Wallingford, Wallingford, UK, 8, 2003.

Ramanathan, V., Crutzen, P. J., Kiehl, J. T., and Rosenfeld, D.: Aerosols, Climate, and the Hydrological Cycle, Science, 294, 2119-2124, 2001.

Renner, M., Werner, M. G. F., Rademacher, S., and Sprokkereef, E.: Verification of ensemble flow forecasts for the River Rhine, Journal of Hydrology, 376, 463-475, doi: http://dx.doi.org/10.1016/j.jhydrol.2009.07.059, 2009.

Rennó, C. D., Nobre, A. D., Cuartas, L. A., Soares, J. V., Hodnett, M. G., Tomasella, J., and Waterloo, M. J.: HAND, a new terrain descriptor using SRTM-DEM: Mapping terra-firme rainforest environments in Amazonia, Remote Sensing of Environment, 112, 3469-3481, doi: http://dx.doi.org/10.1016/j.rse.2008.03.018, 2008.

Richard, Y., Fauchereau, N., Poccard, I., Rouault, M., and Trzaska, S.: 20th century droughts in southern Africa: spatial and temporal variability, teleconnections with oceanic and atmospheric conditions, International Journal of Climatology, 21, 873-885, doi: 10.1002/joc.656, 2001.

RIMES, Regional Integrated Multi-Hazard Early Warning System for Africa and Asia, Water Related Hazard - Drought: http://www.rimes.int/wrh/drought, access: 11 December 2014, 2014.

Robertson, D. E., and Wang, Q. J.: A Bayesian Approach to Predictor Selection for Seasonal Streamflow Forecasting, Journal of Hydrometeorology, 13, 155-171, doi: 10.1175/jhm-d-10-05009.1, 2011.

Rossow, W. B., and Schiffer, R. A.: Advances in Understanding Clouds from ISCCP, Bulletin of the American Meteorological Society, 80, 2261-2287, 1999.

Rouault, M., and Richard, Y.: Intensity and spatial extent of droughts in southern Africa, Geophysical research letters, 32, L15702, doi: 10.1029/2005gl022436, 2005.

Roulin, E.: Skill and relative economic value of medium-range hydrological ensemble predictions, Hydrol. Earth Syst. Sci., 11, 725-737, doi: 10.5194/hess-11-725-2007, 2007.

Savenije, H. H. G.: HESS Opinions "Topography driven conceptual modelling (FLEX-Topo)", Hydrol. Earth Syst. Sci., 14, 2681-2692, doi: 10.5194/hess-14-2681-2010, 2010.

Schulze, R. E.: Hydrological simulation as a tool for agricultural drought assessment, Water S. A., 10, 55-62, 1984.

Schuol, J., and Abbaspour, K. C.: Calibration and uncertainty issues of a hydrological model (SWAT) applied to West Africa, Advances in Geosciences, 9, 7, 2006.

Schuol, J., Abbaspour, K. C., Yang, H., Srinivasan, R., and Zehnder, A. J. B.: Modeling blue and green water availability in Africa, Water Resour. Res., 44, W07406, 2008.

Seibert, M., and Trambauer, P.: Seasonal forecasts of hydrological drought in the Limpopo basin: getting the most out of a bouquet of methods, in: Drought: Research and Science-Policy Interfacing, edited by: Andreu, J., Solera, A., Paredes-Arquiola, J., Haro-Monteagudo, D., and van Lanen, H. A. J., CRC Press, Taylor and Francis Group, 307–313, 2015.

Shanahan, T. M., Overpeck, J. T., Anchukaitis, K. J., Beck, J. W., Cole, J. E., Dettman, D. L., Peck, J. A., Scholz, C. A., and King, J. W.: Atlantic Forcing of Persistent Drought in West Africa, Science, 324, 377-380, doi: 10.1126/science.1166352, 2009.

Sheffield, J., and Wood, E. F.: Projected changes in drought occurrence under future global warming from multi-model, multi-scenario, IPCC AR4 simulations, Climate Dynamics, 31, 79-105, 2008.

Sheffield, J., Wood, E. F., Chaney, N., Guan, K., Sadri, S., Yuan, X., Olang, L., Amani, A., Ali, A., Demuth, S., and Ogallo, L.: A drought monitoring and forecasting system for sub-Sahara African water resources and food security, Bulletin of the American Meteorological Society, 95, 861-882, doi: 10.1175/bams-d-12-00124.1, 2014.

Shukla, S., and Wood, A. W.: Use of a standardized runoff index for characterizing hydrologic drought, Geophysical research letters, 35, L02405, doi: 10.1029/2007gl032487, 2008.

Shukla, S., Sheffield, J., Wood, E. F., and Lettenmaier, D. P.: On the sources of global land surface hydrologic predictability, Hydrol. Earth Syst. Sci., 17, 2781-2796, doi: 10.5194/hess-17-2781-2013, 2013.

Shuttleworth, W. J., Yang, Z. L., and Arain, M. A.: Aggregation rules for surface parameters in global models, Hydrol. Earth Syst. Sci., 1, 217-226, doi: 10.5194/hess-1-217-1997, 1997.

Siebert, S., Döll, P., Feick, S., Hoogeveen, J., and Frenken , K.: Global Map of Irrigation Areas version 4.0.1, ed: Johann Wolfgang Goethe University, Frankfurt am Main, Germany / Food and Agriculture Organization of the United Nations, Rome, Italy, 2007.

Sperna Weiland, F. C., van Beek, L. P. H., Kwadijk, J. C. J., and Bierkens, M. F. P.: The ability of a GCM-forced hydrological model to reproduce global discharge variability, Hydrology and Earth System Sciences, 7, 37, 2010.

Sperna Weiland, F. C., van Beek, L. P. H., Kwadijk, J. C. J., and Bierkens, M. F. P.: On the Suitability of GCM Runoff Fields for River Discharge Modeling: A Case Study Using Model Output from HadGEM2 and ECHAM5, Journal of Hydrometeorology, 13, 140-154, doi: 10.1175/jhm-d-10-05011.1, 2011.

Sperna Weiland, F. C., Tisseuil, C., Dürr, H. H., Vrac, M., and van Beek, L. P. H.: Selecting the optimal method to calculate daily global reference potential evaporation from CFSR reanalysis data for application in a hydrological model study, Hydrol. Earth Syst. Sci., 16, 983-1000, 2012.

Stewart, J. B., Engman, E. T., Feddes, R. A., and Kerr, Y.: Scaling up in hydrology using remote sensing, John Wiley and Sons, 255 pp., 1996.

Sylla, M. B., Coppola, E., Mariotti, L., Giorgi, F., Ruti, P., Dell'Aquila, A., and Bi, X.: Multiyear simulation of the African climate using a regional climate model (RegCM3) with the high resolution ERA-interim reanalysis, Climate Dynamics, 35, 231-247, 2010.

Szczypta, C., Calvet, J. C., Albergel, C., Balsamo, G., Boussetta, S., Carrer, D., Lafont, S., and Meurey, C.: Verification of the new ECMWF ERA-Interim reanalysis over France, Hydrology and Earth System Sciences, 15, 647-666, 2011.

Takata, K., Emori, S., and Watanabe, T.: Development of the minimal advanced treatments of surface interaction and runoff, Global and Planetary Change, 38, 209-222, 2003.

Tallaksen, L. M., and Van Lanen, H. A. J.: Hydrological drought: processes and estimation methods for streamflow and groundwater, Elsevier Science, 2004.

Teuling, A., Hirschi, M., Ohmura, A., Wild, M., Reichstein, M., Ciais, P., Buchmann, N., Ammann, C., Montagnani, L., and Richardson, A.: A regional perspective on trends in continental evaporation, Geophys. Res. Lett, 36, L02404, doi:10.1029/2008GL036584, 2009.

Teuling, A. J., Van Loon, A. F., Seneviratne, S. I., Lehner, I., Aubinet, M., Heinesch, B., Bernhofer, C., Grünwald, T., Prasse, H., and Spank, U.: Evapotranspiration amplifies European summer drought, Geophysical research letters, 40, 2071-2075, doi: 10.1002/grl.50495, 2013.

Thiemig, V., Rojas, R., Zambrano-Bigiarini, M., Levizzani, V., and De Roo, A.: Validation of Satellite-Based Precipitation Products Over Sparsely-Gauged African River Basins, Journal of Hydrometeorology, 13, 1760–1783, doi: 10.1175/JHM-D-12-032.1, 2012.

Thiemig, V., Rojas, R., Zambrano-Bigiarini, M., and De Roo, A.: Hydrological evaluation of satellite-based rainfall estimates over the Volta and Baro-Akobo Basin, Journal of Hydrology, 499, 324-338, doi: http://dx.doi.org/10.1016/j.jhydrol.2013.07.012, 2013.

Thomson, M. C., Abayomi, K., Barnston, A. G., Levy, M., and Dilley, M.: El Niño and drought in southern Africa, The Lancet, 361, 437-438, 2003.

Tierney, J. E., Smerdon, J. E., Anchukaitis, K. J., and Seager, R.: Multidecadal variability in East African hydroclimate controlled by the Indian Ocean, Nature, 493, 389-392, doi: http://www.nature.com/nature/journal/v493/n7432/abs/nature11785.html#supplementary-information, 2013.

Touchan, R., Anchukaitis, K. J., Meko, D. M., Attalah, S., Baisan, C., and Aloui, A.: Long term context for recent drought in northwestern Africa, Geophysical research letters, 35, L13705, doi: 10.1029/2008gl034264, 2008.

Touchan, R., Anchukaitis, K., Meko, D., Sabir, M., Attalah, S., and Aloui, A.: Spatiotemporal drought variability in northwestern Africa over the last nine centuries, Climate Dynamics, 37, 237-252, doi: 10.1007/s00382-010-0804-4, 2011.

Trambauer, P., Maskey, S., Winsemius, H., Werner, M., and Uhlenbrook, S.: A review of continental scale hydrological models and their suitability for drought forecasting in (sub-Saharan) Africa, Physics and Chemistry of the Earth, 66, 16-26, doi: http://dx.doi.org/10.1016/j.pce.2013.07.003, 2013.

Trambauer, P., Dutra, E., Maskey, S., Werner, M., Pappenberger, F., van Beek, L. P. H., and Uhlenbrook, S.: Comparison of different evaporation estimates over the African continent, Hydrol. Earth Syst. Sci., 18, 193-212, doi: 10.5194/hess-18-193-2014, 2014a.

Trambauer, P., Maskey, S., Werner, M., Pappenberger, F., van Beek, L. P. H., and Uhlenbrook, S.: Identification and simulation of space–time variability of past hydrological drought events in the Limpopo River basin, southern Africa, Hydrol. Earth Syst. Sci., 18, 2925-2942, doi: 10.5194/hess-18-2925-2014, 2014b.

Trambauer, P., Werner, M., Winsemius, H. C., Maskey, S., Dutra, E., and Uhlenbrook, S.: Hydrological drought forecasting and skill assessment for the Limpopo River basin, southern Africa, Hydrol. Earth Syst. Sci., 19, 1695-1711, doi: 10.5194/hess-19-1695-2015, 2015.

Trenberth, K.: Framing the way to relate climate extremes to climate change, Climatic Change, 115, 283-290, doi: 10.1007/s10584-012-0441-5, 2012.

Tveito, O. E., Wegehenkel, M., van der Wel, F., and Dobesch, H.: COST Action 719: The Use of Geographic Information Systems in Climatology and Meteorology: Final Report, EUR-OP, Luxembourg: Office for Official Publications of the European Communities, 254 pp, 2008.

UN: Department of International Economic Social Affairs United Nations, World population to 2300, United Nations, New York, http://www.un.org/esa/population/publications/longrange2/WorldPop2300final.pdf, access: 27 October 2014, 2004.

UNCCD, FAO, and WMO: High Level Meeting on National Drought Policy (HMNDP) CICG, Geneva, 11-15 March 2013, Science Document: Best Practices on National Drought Management Policy, http://www.hmndp.org/sites/default/files/docs/ScienceDocument14212_Eng.pdf, access: 05/12/2014, 2012.

UNDP: A global report: reducing disaster risk, a challenge for development., United Nations Development Programme, http://www.undp.org/content/dam/undp/library/crisis%20prevention/disaster/asia_pacific/Reduc ing%20Disaster%20risk%20a%20Challenge%20for%20development.pdf, 146, 2004.

UNECA-ACPC: Management of Ground Water in Africa Including Transboundary Aquifers: Implications for Food Security, Livelihood and Climate Change Adaptation, United Nations Economic Commission for Africa - African Climate Policy Centre, http://www.uneca.org/sites/default/files/publications/wp6-groundwater_final_draft.pdf, access 25/03/2015, 2011.

UNEP: World atlas of desertification 2ED, UNEP (United Nations Environment Programme) London, 1997.

UNISDR: International strategy for disaster reduction "Hyogo framework for action 2005–2015: Building the Resilience of Nations and Communities to Disasters", World conference on disaster reduction, available at: http://www.unisdr.org/files/1037_hyogoframeworkforactionenglish.pdf, last access: 28 March 2015, Kobe, Japan, 28 pp, 2005,

UNISDR, International Strategy for Disaster Reduction - Platform for the Promotion of Early Warning, Basics of early warning: http://www.unisdr.org/2006/ppew/whats-ew/basics-ew.htm, access: 22 October 2014, 2006.

University of New Hampshire, Water Systems Analysis Group, The WBM Model Family: http://www.wsag.unh.edu/wbm2.html, access: May 2011, 2009.

USGS, MODIS Overview https://lpdaac.usgs.gov/products/modis_overview, access: April 2012, 2012.

USGS EROS, Africa Land Cover Characteristics Data Base Version 2.0: http://edc2.usgs.gov/glcc/tablambert_af.php, access: 24 June 2012, 2002.

USGS EROS, Hydro1K Africa http://eros.usgs.gov/#/Find_Data/Products_and_Data_Available/gtopo30/hydro/africa, access: 21 September 2012, 2006.

van Beek, L. P. H.: Forcing PCR-GLOBWB with CRU data, Utrecht University, Utrecht, Netherlands: http://vanbeek.geo.uu.nl/suppinfo/vanbeek2008.pdf, last access: December 2013, 2008.

van Beek, L. P. H., and Bierkens, M. F. P.: The Global Hydrological Model PCR-GLOBWB: Conceptualization, Parameterization and Verification, Utrecht University, Faculty of Earth Sciences, Department of Physical Geography, Utrecht, The Netherlands, available at: http://vanbeek.geo.uu.nl/suppinfo/vanbeekbierkens2009.pdf, last access: 28 March 2015, 53 pp, 2009.

van Beek, L. P. H., Wada, Y., and Bierkens, M. F. P.: Global monthly water stress: 1. Water balance and water availability, Water Resour. Res., 47, W07517, doi: 10.1029/2010WR009791, 2011.

Van den Hurk, B., Viterbo, P., Beljaars, A. C. M., and Betts, A. K.: Offline validation of the ERA40 surface scheme, ECMWF Tech. Memo. 295 1-42, 2000.

van der Ent, R. J., Savenije, H. H. G., Schaefli, B., and Steele-Dunne, S. C.: Origin and fate of atmospheric moisture over continents, Water Resources Research, 46, W09525, doi: 10.1029/2010WR009127, 2010.

van der Knijff, J., and de Roo, A. P. J.: LISFLOOD - Distributed Water Balance and Flood Simulation Model - Revised User Manual (published on http://publications.jrc.ec.europa.eu/repository/handle/111111111/7776), JRC Scientific and Technical Reports 2008.

Velázquez, J. A., Anctil, F., Ramos, M. H., and Perrin, C.: Can a multi-model approach improve hydrological ensemble forecasting? A study on 29 French catchments using 16 hydrological model structures, Adv. Geosci., 29, 33-42, doi: 10.5194/adgeo-29-33-2011, 2011.

Verkade, J. S., Brown, J. D., Reggiani, P., and Weerts, A. H.: Post-processing ECMWF precipitation and temperature ensemble reforecasts for operational hydrologic forecasting at various spatial scales, Journal of Hydrology, 501, 73-91, doi: http://dx.doi.org/10.1016/j.jhydrol.2013.07.039, 2013.

Verschuren, D., Laird, K. R., and Cumming, B. F.: Rainfall and drought in equatorial east Africa during the past 1,100 years, Nature, 403, 410-414, doi: http://www.nature.com/nature/journal/v403/n6768/suppinfo/403410a0_S1.html, 2000.

Vicente-Serrano, S. M., Beguería, S., and López-Moreno, J. I.: A Multiscalar Drought Index Sensitive to Global Warming: The Standardized Precipitation Evapotranspiration Index, Journal of Climate, 23, 1696-1718, doi: 10.1175/2009jcli2909.1, 2010a.

Vicente-Serrano, S. M., Beguería, S., López-Moreno, J. I., Angulo, M., and El Kenawy, A.: A New Global 0.5° Gridded Dataset (1901–2006) of a Multiscalar Drought Index: Comparison with Current Drought Index Datasets Based on the Palmer Drought Severity Index, Journal of Hydrometeorology, 11, 1033-1043, 2010b.

Vicente-Serrano, S. M., Beguería, S., Gimeno, L., Eklundh, L., Giuliani, G., Weston, D., El Kenawy, A., López-Moreno, J. I., Nieto, R., Ayenew, T., Konte, D., Ardö, J., and Pegram, G. G. S.: Challenges for drought mitigation in Africa: The potential use of geospatial data and drought information systems, Applied Geography, 34, 471-486, doi: http://dx.doi.org/10.1016/j.apgeog.2012.02.001, 2012.

Viney, N. R., and Sivapalan, M.: A framework for scaling of hydrologic conceptualizations based on a disaggregation–aggregation approach, Hydrological Processes, 18, 1395-1408, doi: 10.1002/hyp.1419, 2004.

Vinukollu, R. K., Meynadier, R., Sheffield, J., and Wood, E. F.: Multi-model, multi-sensor estimates of global evapotranspiration: climatology, uncertainties and trends, Hydrological Processes, 25, 3993–4010, doi: 10.1002/hyp.8393, 2011.

Viterbo, P., Beljaars, A., Mahfouf, J.-F., and Teixeira, J.: The representation of soil moisture freezing and its impact on the stable boundary layer, Quarterly Journal of the Royal Meteorological Society, 125, 2401-2426, 1999.

Vörösmarty, C. J., Moore, B., III, Grace, A. L., Gildea, M. P., Melillo, J. M., Peterson, B. J., Rastetter, E. B., and Steudler, P. A.: Continental scale models of water balance and fluvial transport: An application to South America, Global Biogeochem. Cycles, 3, 241-265, 1989.

Voß, F., and Alcamo, J.: Technical Report No. 1: First results from intercomparison of surface water availability modules (published on: http://www.eu-watch.org/), WATCH, 2008.

Wada, Y., van Beek, L. P. H., van Kempen, C. M., Reckman, J. W. T. M., Vasak, S., and Bierkens, M. F. P.: Global depletion of groundwater resources, Geophysical research letters, 37, L20402, doi: 10.1029/2010gl044571, 2010.

Wada, Y., van Beek, L. P. H., Viviroli, D., Dürr, H. H., Weingartner, R., and Bierkens, M. F. P.: Global monthly water stress: 2. Water demand and severity of water stress, Water Resources Research, 47, W07518, doi: 10.1029/2010wr009792, 2011.

Wanders, N., Lanen, H. A. J., and van Loon, A. F.: Indicators for Drought Characterization on a Global Scale. WATCH Water and Global change EU FP6. Technical report No. 24, http://www.eu-watch.org/publications/technical-reports/3, last access: December 2013, 2010.

Wang, L., Ranasinghe, R., Maskey, S., van Gelder, P. H. A. J. M., and Vrijling, J. K.: Comparison of empirical statistical methods for downscaling daily climate projections from CMIP5 GCMs: a case study of the Huai River Basin, China, Int. J. Climatol., doi: 10.1002/joc.4334, 2015.

Wang, N.-Y., Liu, C., Ferraro, R., Wolff, D., Zipser, E., and Kummerow, C.: TRMM 2A12 Land Precipitation Product - Status and Future Plans, Journal of the Meteorological Society of Japan. Ser. II, 87A, 237-253, 2009.

Wang, T., and Zlotnik, V. A.: A complementary relationship between actual and potential evapotranspiration and soil effects, Journal of Hydrology, 456–457, 146-150, doi: http://dx.doi.org/10.1016/j.jhydrol.2012.03.034, 2012.

Water Research Commission: IWR Rhodes University, School of BEEH University of KwaZulu-Natal, and Water for Africa: Identification, estimation, quantification and incorporation of risk and uncertainty in water resources management tools in South Africa. Deliverable No. 3: Interim Report on Sources of Uncertainty, Water Research Commission Project No: K5/1838 http://www.ru.ac.za/static/institutes/iwr/uncertainty/k51838/DEL3_Uncertainty_Sources.pdf (last access: August 2014), 2009.

Weiß, M., and Menzel, L.: A global comparison of four potential evapotranspiration equations and their relevance to stream flow modelling in semi-arid environments, Adv. Geosci., 18, 15-23, 2008.

Werner, K., Brandon, D., Clark, M., and Gangopadhyay, S.: Climate Index Weighting Schemes for NWS ESP-Based Seasonal Volume Forecasts, Journal of Hydrometeorology, 5, 1076-1090, doi: 10.1175/jhm-381.1, 2004.

Werner, M., Schellekens, J., Gijsbers, P., van Dijk, M., van den Akker, O., and Heynert, K.: The Delft-FEWS flow forecasting system, Environmental Modelling & Software, 40, 65-77, doi: http://dx.doi.org/10.1016/j.envsoft.2012.07.010, 2013.

Werner, M., Vermooten, S., Iglesias, A., Maia, R., Vogt, J., and Naumann, G.: Developing a framework for drought forecasting and warning: Results of the DEWFORA project, in: Drought: Research and Science-Policy Interfacing, CRC Press, Taylor and Francis Group, 279-285, 2015.

West, L., What are the Effects of Drought?: http://environment.about.com/od/environmentalevents/a/droughteffects.htm, access: 20 November 2014, 2014.

Western, A. W., Grayson, R. B., Blöschl, G., Willgoose, G. R., and McMahon, T. A.: Observed spatial organization of soil moisture and its relation to terrain indices, Water Resources Research, 35, 797-810, doi: 10.1029/1998wr900065, 1999.

Widén-Nilsson, E., Halldin, S., and Xu, C.-y.: Global water-balance modelling with WASMOD-M: Parameter estimation and regionalisation, Journal of Hydrology, 340, 105-118, 2007.

Widén-Nilsson, E., Gong, L., Halldin, S., and Xu, C. Y.: Model performance and parameter behavior for varying time aggregations and evaluation criteria in the WASMOD-M global water balance model, Water Resour. Res., 45, 14, 2009.

Wigmosta, M., and Prasad, R.: Upscaling and Downscaling – Dynamic Models, in: Encyclopedia of Hydrological Sciences, John Wiley & Sons, Ltd, 2006.

Wilhite, D. A., and Glantz, M. H.: Understanding: the Drought Phenomenon: The Role of Definitions, Water International, 10, 111-120, doi: 10.1080/02508068508686328, 1985.

Wilks, D. S.: Statistical methods in the atmospheric sciences - 3rd edition, International Geophysics series, vol 100, 676 pp., 2011.

Williams, A., and Funk, C.: A westward extension of the warm pool leads to a westward extension of the Walker circulation, drying eastern Africa, Climate Dynamics, 1-19, 2011.

Winsemius, H. C., Dutra, E., Engelbrecht, F. A., Archer Van Garderen, E., Wetterhall, F., Pappenberger, F., and Werner, M. G. F.: The potential value of seasonal forecasts in a changing climate in southern Africa, Hydrol. Earth Syst. Sci., 18, 1525-1538, doi: 10.5194/hess-18-1525-2014, 2014.

Wipfler, E. L., Metselaar, K., van Dam, J. C., Feddes, R. A., van Meijgaard, E., van Ulft, L. H., van den Hurk, B., Zwart, S. J., and Bastiaanssen, W. G. M.: Seasonal evaluation of the land surface scheme HTESSEL against remote sensing derived energy fluxes of the Transdanubian region in Hungary, Hydrol. Earth Syst. Sci., 15, 1257-1271, 2011.

Wisser, D., Fekete, B. M., Vörösmarty, C. J., and Schumann, A. H.: Reconstructing 20th century global hydrography: a contribution to the Global Terrestrial Network- Hydrology (GTN-H), Hydrol. Earth Syst. Sci., 14, 1-24, 2010.

WMO: Limpopo River basin - A proposal to improve the flood forecasting and early warning system, World Meteorological Organization, http://www.wmo.int/pages/prog/hwrp/chy/chy14/documents/ms/Limpopo_Report.pdf, last access: December 2013, 2012a.

WMO: Guidelines on Ensemble prediction systems and forecasting, World Meteorological Organization, http://www.wmo.int/pages/prog/www/Documents/1091_en.pdf, last access: 27 October 2014, 2012b.

Wood, A. W., and Lettenmaier, D. P.: An ensemble approach for attribution of hydrologic prediction uncertainty, Geophysical research letters, 35, L14401, doi: 10.1029/2008gl034648, 2008.

Wood, E. F., Roundy, J. K., Troy, T. J., van Beek, L. P. H., Bierkens, M. F. P., Blyth, E., de Roo, A., Döll, P., Ek, M., Famiglietti, J., Gochis, D., van de Giesen, N., Houser, P., Jaffé, P. R., Kollet, S., Lehner, B., Lettenmaier, D. P., Peters-Lidard, C., Sivapalan, M., Sheffield, J., Wade, A., and Whitehead, P.: Hyperresolution global land surface modeling: Meeting a grand challenge for monitoring Earth's terrestrial water, Water Resources Research, 47, W05301, doi: 10.1029/2010wr010090, 2011.

WWRP/WGNE, Forecast verification: Issues, Methods and FAQ. Joint Working Group on Verification sponsored by the World Meteorological Organization, available at: http://www.cawcr.gov.au/projects/verification/#Methods_for_probabilistic_forecasts, access: 26 August 2014, 2013.

Xia, Y.: Calibration of LaD model in the northeast United States using observed annual streamflow, Journal of Hydrometeorology, 8, 1098-1110, 2007.

Yamazaki, D., Kanae, S., Kim, H., and Oki, T.: A physically based description of floodplain inundation dynamics in a global river routing model, Water Resources Research, 47, 21, 2011.

Yossef, N. C., van Beek, L. P. H., Kwadijk, J. C. J., and Bierkens, M. F. P.: Skill assessment of a global hydrological model in reproducing flow extremes, Hydrol. Earth Syst. Sci. Discuss., 8, 3469-3505, doi:10.5194/hessd-8-3469-2011, 2011.

Yossef, N. C., Winsemius, H., Weerts, A., van Beek, R., and Bierkens, M. F. P.: Skill of a global seasonal streamflow forecasting system, relative roles of initial conditions and meteorological forcing, Water Resources Research, 49, 4687-4699, doi: 10.1002/wrcr.20350, 2013.

Yuan, X., and Wood, E. F.: Downscaling precipitation or bias-correcting streamflow? Some implications for coupled general circulation model (CGCM)-based ensemble seasonal hydrologic forecast, Water Resources Research, 48, W12519, doi: 10.1029/2012wr012256, 2012.

Yuan, X., Wood, E. F., Chaney, N. W., Sheffield, J., Kam, J., Liang, M., and Guan, K.: Probabilistic Seasonal Forecasting of African Drought by Dynamical Models, Journal of Hydrometeorology, 14, 1706-1720, doi: 10.1175/jhm-d-13-054.1, 2013.

Zalachori, I., Ramos, M. H., Garçon, R., Mathevet, T., and Gailhard, J.: Statistical processing of forecasts for hydrological ensemble prediction: a comparative study of different bias correction strategies, Adv. Sci. Res., 8, 135-141, doi: 10.5194/asr-8-135-2012, 2012.

Zargar, A., Sadiq, R., Naser, B., and Khan, F. I.: A review of drought indices, Environmental Reviews, 19, 333-349, doi:10.1139/A11-013, 2011.

Zeng, N.: Drought in the Sahel, Science, 302, 999-1000, doi: 10.1126/science.1090849, 2003.

Zhang, K., Kimball, J. S., Nemani, R. R., and Running, S. W.: A continuous satellite-derived global record of land surface evapotranspiration from 1983 to 2006, Water Resour. Res, 46, W09522, doi:10.1029/2009WR008800, 2010.

Zhao, M., Running, S. W., and Nemani, R. R.: Sensitivity of Moderate Resolution Imaging Spectroradiometer (MODIS) terrestrial primary production to the accuracy of meteorological reanalyses, J. Geophys. Res., 111, G01002, doi: 10.1029/2004jg000004, 2006.

Zhu, T., and Ringler, C.: Climate Change Impacts on Water Availability and Use in the Limpopo River Basin, Water, 4, 63-84, 2012.

Zhu, Y.: Ensemble forecast: A new approach to uncertainty and predictability, Advances in atmospheric sciences, 22, 781-788, 2005.

Zomer RJ, Trabucco A, Bossio DA, van Straaten O, and Verchot LV Climate Change Mitigation: A Spatial Analysis of Global Land Suitability for Clean Development Mechanism Afforestation and Reforestation, Agric. Ecosystems and Envir., 126 67-80, 2008.

Acronyms

ACRU	Agricultural Catchments Research Unit
ADO	African Drought Observatory
AI	Aridity Index [-]
AIRS	Atmospheric InfraRed Sounder
AMS	American Meteorological Society
BS	Brier Score [-]
BSS	Brier Skill Score [-]
CFSv2	NCEP's Climate Forecast System version 2
CGIAR-CSI	Consultative Group for International Agriculture Research Consortium for Spatial Information
CMI	Crop Moisture Index [-]
CSIR	Council for Scientific and Industrial Research
CV	Coefficient of variation [-]
DEM	Digital Elevation Model
DEWFORA	Improved Drought Early Warning and FORecasting to strengthen preparedness and adaptation to droughts in Africa
DEWS	Drought Early Warning Systems
DS	Drought Severity [months]
ECMWF	European Centre for Medium-Range Weather Forecasts
EDO	European Drought Observatory
EM	Evaporation Multiproduct
EM-DAT	Emergency Event Database
ENSO	El Niño Southern Oscillation
ERA-40	40-year ECMWF reanalysis
ERAI	ERA-Interim ECMWF reanalysis
ERAL	ERA-Land
ESP	Ensemble Streamflow Prediction
ESPcond	Ensemble Streamflow Prediction conditional on the ENSO signal
ETDI	Evapotranspiration Deficit Index [-]
EVI	Enhanced Vegetation Index [-]
EWS	Early Warning Systems
FAO	Food and Agriculture Organization of the United Nations
FEWS Net	Famine Early Warning Systems Network
FS	Forecasting System
FS_S4	Hydrological Forecasting System forced with ECMWF seasonal forecast system S4

FS_ESP	Hydrological Forecasting System with the ESP approach
FS_ESPcond	Hydrological Forecasting System with the ESPcond approach
GAI	Global Aridity Index
GCMs	General Circulation Models
GDEWF	Global Drought Early Warning Monitoring Framework
GDEWS	Global Drought Early Warning System
GEWEX	Global Energy and Water Exchanges Project
GHMs	Global Hydrological Models
GIEWS	FAO's Global Information and Early Warning System on Food and Agriculture
GLCC	Global Land Cover Characterization
GLDAS	Global Land Data Assimilation System
GLEAM	Global Land surface Evaporation: the Amsterdam Methodology
GLWD	Global Lakes and Wetlands Database
GMAO	NASA's Global Modelling and Assimilation Office
GPCP	Global Precipitation Climatology Project
GRDC	Global Runoff Data Center
GRI	Groundwater Resource Index [-]
GWAVA	Global Water Availability Assessment method
HAND	Height above the nearest drainage [m]
HEPEX	Hydrological Ensemble Prediction Experiment
HEWS	Humanitarian Early Warning Service
HTESSEL	Hydrology Tiled ECMWF Scheme for Surface Exchanges over Land
IHCs	Initial hydrologic conditions
IPCC	Intergovernmental Panel on Climate Change
ISCCP	International Satellite Cloud Climatology Project
JRC	European Commission Joint Research Centre
KGE	Kling-Gupta Efficiency [-]
LaD	Land Dynamics Model
LAI	Leaf Area Index [m m^{-1}]
LPJ	Lund-Postdam-Jena model
LPRM	Land Parameter Retrieval Model
LSA-SAF	Land Surface Analysis Satellite Application Facility
LSMs	Land Surface Models
Mac-PDM	Macro-scale-Probability-Distributed Moisture Model
MAP	Mean Annual Precipitation [mm yr^{-1}]
MAPE	Mean Annual Potential Evaporation [mm yr^{-1}]
MATSIRO	Minimal Advanced Treatments of Surface Interaction and Runoff
MF	Meteorological Forecast
MODIS	MODerate resolution Image Spectroradiometer
MOD16	Potential and actual evaporation product from MODIS
NASA	National Aeronautics and Space Administration
NCEP	National Center for Environmental Prediction's

NOAA	National Oceanic and Atmospheric Administration
NMME	North American Multi-Model Ensemble
NSE	Nash-Sutcliffe Efficiency [-]
NSIDC	National Snow and Ice Data Center
ONI	Oceanic Niño Index
PBIAS	Percent Bias [-]
PCR-GLOBWB	PCRaster GLOBal Water Balance model
PCR_Irrig	PCR-GLOBWB with an irrigation module
PCR_PM	PCR-GLOBWB forced with Penman-Monteith potential evaporation
PCR_TRMM	PCR-GLOBWB forced with TRMM 3B42 v6 precipitation data
PDM	Probability Distributed Moisture Model
PDSI	Palmer Drought Severity Index [-]
PERSIANN	Precipitation Estimation from Remotely Sensed Information using Artificial Neural Networks
PET_c	Crop-specific potential evaporation [mm d^{-1}]
PET_r	Reference potential evaporation [mm d^{-1}]
PHDI	Palmer Hydrological Drought Index [-]
POD	Probability of Detection [-]
POFD	Probability of False Detection [-]
R^2	Coefficient of determination [-]
RC	Runoff Coefficient [-]
RIMES	Regional Integrated Multi-Hazard Early Warning System for Africa and Asia
RMB	Relative Mean Bias [%]
RMD	Relative Mean Difference [%]
RMSE	Root Mean Square Error
ROC	Relative Operating Characteristic [-]
ROCS	ROC Score [-]
RS	Root Stress [-]
RSA	Root Stress Anomaly [-]
RSAI	Root Stress Anomaly Index [-]
RSR	Ratio of the root mean square error to the standard deviation [-]
SARCOF	Southern Africa Regional Climate Outlook Forum
SPI	Standardized Precipitation Index [-]
SPEI	Standardized Precipitation Evaporation Index [-]
SRI	Standardized Runoff Index [-]
SWAT	Soil and Water Assessment Tool
TRMM	Tropical Rainfall Measuring Mission
UNCCD	UN Convention for Combating Desertification
UNDP	United Nations Development Programme
UNISDR	United Nations Office for Disaster Risk Reduction
USDA	US Department of Agriculture
VIC	Variable Infiltration Capacity
WBM	Water Balance Model

WFP	World Food Programme
WMO	World Meteorological Organization
WSA	Water stress anomaly [-]

List of publications

Journal papers

Trambauer P., Maskey S., Winsemius H., Werner M., and Uhlenbrook S.: A review of continental scale hydrological models and their suitability for drought forecasting in (sub-Saharan) Africa, Physics and Chemistry of the Earth, 66, 16-26, doi: http://dx.doi.org/10.1016/j.pce.2013.07.003, 2013.

Trambauer P., Dutra E., Maskey S., Werner M., Pappenberger F., van Beek L. P. H., and Uhlenbrook S.: Comparison of different evaporation estimates over the African continent, Hydrol. Earth Syst. Sci., 18, 193-212, doi: 10.5194/hess-18-193-2014, 2014.

Trambauer P., Maskey S., Werner M., Pappenberger F., van Beek L. P. H., and Uhlenbrook S.: Identification and simulation of space-time variability of past hydrological drought events in the Limpopo river basin, southern Africa, Hydrol. Earth Syst. Sci., 18, 2925-2942, doi: 10.5194/hess-18-2925-2014, 2014.

Masih I., Maskey S., Mussá F.E.F., and Trambauer P.: A review of droughts on the African continent: a geospatial and long-term perspective, Hydrol. Earth Syst. Sci., 18, 3635-3649, doi: 10.5194/hess-18-3635-2014, 2014.

Trambauer, P., Werner, M., Winsemius, H. C., Maskey, S., Dutra, E., and Uhlenbrook, S.: Hydrological drought forecasting and skill assessment for the Limpopo River basin, southern Africa, Hydrol. Earth Syst. Sci., 19, 1695-1711, doi:10.5194/hess-19-1695-2015, 2015.

Trambauer, P., Maskey, S., Werner, M., Uhlenbrook, S. and van Beek, L.P.H: Downscaling the output of a low resolution hydrological model to higher resolutions, submitted to Hydrological Processes.

Book chapters

Maskey, S. and Trambauer, P.: Hydrological modeling for drought assessment, in: Hydro-Meteorological Hazards, Risks, and Disasters, edited by: Shroder, J. F., Paron, P., and Di Baldassarre, G., Elsevier, 2014.

Seibert, M. and Trambauer, P.: Seasonal forecasts of hydrological drought in the Limpopo basin: Getting the most out of a bouquet of methods, in: Drought: Research and Science-Policy Interfacing, edited by Andreu, J., Solera, A., Paredes-Arquiola, J., Haro-Monteagudo, D., and van

Lanen, H., CRC Press, Taylor and Francis Group, pp. 307–313, ISBN: 978-1-138-02779-4, doi: 10.1201/b18077-52, 2015.

Conference presentations and posters

Trambauer P., Maskey S., Winsemius H., Werner M., and Uhlenbrook S. A review of continental scale hydrological models and their suitability for drought forecasting in (sub-Saharan) Africa. 12th WaterNet/WARFSA/GWP-SA Symposium, Maputo, Mozambique, October 2011 [poster presentation].

Trambauer P., Maskey S., Werner M., and Uhlenbrook S. Identifying historic droughts in the Limpopo river basin using a downscaled version of PCR-GLOBWB. 13th WaterNet/WARFSA/GWP-SA Symposium, Johannesburg, South Africa, October 2012 [oral presentation].

Trambauer P., Maskey S., and Werner M. A comparison of continental actual evaporation estimates for Africa to improve hydrological drought forecasting. EGU General Assembly, Vienna, Austria, April 2013 [oral presentation; abstract: Geophysical Research Abstracts, Vol. 15, EGU2013-13330].

Maskey S., and **Trambauer P.** Modelling hydrological droughts in a (semi-)arid basin. Procs. 8th International Conference of European Water Resources Association. Porto, Portugal, 26-29 June 2013 [oral presentation by Maskey S.].

Trambauer P., Werner M., and Maskey S. Seasonal hydrological forecasting for drought early warning in the Limpopo river basin. 14th WaterNet/WARFSA/GWP-SA Symposium, Dar es Salaam, Tanzania, October 2013 [oral presentation].

Trambauer P., Werner M., Winsemius H.C., Maskey S., Dutra E., and Uhlenbrook S. Hydrological drought forecasting and skill assessment for the Limpopo river basin, Southern Africa. EGU General Assembly, Vienna, Austria, April-May 2014 [oral presentation; abstract: Geophysical Research Abstracts, Vol. 16, EGU2014-11782].

Werner M., **Trambauer P.,** Winsemius H.C., Maskey S., Dutra E., and Pappenberger F. Assessing the skill of seasonal forecasting of streamflow and drought in the Limpopo basin, southern Africa. AGU Fall Meeting, San Francisco, USA, December 2014 [poster presentation by Werner M., H11F-0920A]

Seibert M., and **Trambauer P.,** Seasonal forecasts of hydrological drought in the Limpopo basin: getting the most out of a bouquet of methods. International Conference on DROUGHT: Research and Science-Policy Interfacing. Valencia, Spain, March 2015 [oral presentation].

Biography

Patricia M. Trambauer Arechavaleta was born in Montevideo, Uruguay, on 30 May 1982. She obtained her B.Sc. in Civil Engineering, specialization Hydraulic and Environmental Engineering in 2007 from the Faculty of Engineering, State University in Montevideo, Uruguay. While studying she was already working at the consultant company CSI Ingenieros as a project engineer assistant, and after graduation she became a project engineer at the same company. Her main responsibilities included the design of sewerage networks, drainage networks, water treatment plants and assisting in environmental impact assessment studies.

In 2010, she obtained her MSc in Water Science and Engineering - specialization Hydrology and Water Resources from UNESCO-IHE, Delft, The Netherlands with distinction. Her thesis was entitled "Surface water and shallow groundwater flow systems in lowland peat areas – Case study at the Zegveld experimental farm". After her MSc studies, she return to CSI Ingenieros in Montevideo and worked for a few months in a project that involved the supply of drinking water to several arid regions in Angola, where the only source of supply water is groundwater.

In February 2011, she started working on her PhD research at the Department of Water Science and Engineering at UNESCO-IHE. Her research was part of a larger project "Improved Drought Early Warning and Forecasting to strengthen preparedness and adaptation in Africa" (DEWFORA), a collaborative project funded under the Seventh Framework Programme (FP-7) of the EU under the theme "Early warning and forecasting systems to predict climate related drought vulnerability and risks in Africa" (ENV 2010.1.3.3-1). The project involved the collaboration of 19 European and African partners; further details can be found at www.dewfora.com.

Netherlands Research School for the
Socio-Economic and Natural Sciences of the Environment

D I P L O M A

For specialised PhD training

The Netherlands Research School for the
Socio-Economic and Natural Sciences of the Environment
(SENSE) declares that

Patricia María Trambauer Arechavaleta

born on 30 May 1982 in Montevideo, Uruguay

has successfully fulfilled all requirements of the
Educational Programme of SENSE.

Delft, 23 June 2015

the Chairman of the SENSE board

Prof. dr. Huub Rijnaarts

the SENSE Director of Education

Dr. Ad van Dommelen

The SENSE Research School declares that Ms Patricia Trambauer has successfully fulfilled all requirements of the Educational PhD Programme of SENSE with a work load of 46.9 EC, including the following activities:

SENSE PhD Courses

o   Environmental Research in Context (2011)
o   Research in Context Activity: 'Co-organising the annual UNESCO-IHE PhD week, including SENSE A1 Course', Delft, The Netherlands (2011)

Other PhD and Advanced MSc Courses

o   Writing a Scientific article, VU University Amsterdam (2012)
o   Summer School 'Drought hazard and Management: Challenges in a Changing world', National Technical University of Athens, Greece (2014)
o   Data Analysis and Statistical Inference - E-learning Coursera Course, Duke University (2014)

Management and Didactic Skills Training

o   Teaching assistant in MSc course 'Engineering Hydrology', UNESCO-IHE, Delft (2011)
o   Organising the annual PhD Seminar, UNESCO-IHE, Delft (2011)
o   Chair of the IHE's PhD fellows Association Board (PAB) (2012-2014)
o   Guest lecturer in advanced course 'Drought forecasting and its use in informed decision making', CIHEAM, Zaragoza, Spain (2013)

Oral Presentations

o   Identifying historic droughts in the Limpopo river basin using a downscaled version of PCR-GLOBWB. 13th WaterNet/WARFSA/GWP-SA Symposium, 31 October-2 November 2012, Johannesburg, South Africa
o   A comparison of different continental actual evaporation estimates for Africa to improve hydrological drought forecasting from modelled indicators. European Geosciences Union (EGU) General Assembly, 7-12 April 2013, Vienna, Austria
o   *Seasonal hydrological forecasting for drought early warning in the Limpopo river basin.* 14th WaterNet/WARFSA/GWP-SA Symposium, 30 October-1 November 2013, Dar es Salaam, Tanzania
o   *Hydrological drought forecasting and skill assessment for the Limpopo river basin, Southern Africa.* European Geosciences Union (EGU) General Assembly, 27 April-2 May 2014, Vienna, Austria

SENSE Coordinator PhD Education

Dr. ing. Monique Gulickx

Printed and bound by CPI Group (UK) Ltd, Croydon, CR0 4YY

21/10/2024

01777101-0005